History of Mathematics · *Volume 39*

SOURCES

THEORY OF ALGEBRAIC FUNCTIONS OF ONE VARIABLE

RICHARD DEDEKIND

HEINRICH WEBER

TRANSLATED AND INTRODUCED BY JOHN STILLWELL

American Mathematical Society
London Mathematical Society

2010 *Mathematics Subject Classification.* Primary 01-02, 01A55.

For additional information and updates on this book, visit
www.ams.org/bookpages/hmath-39

Library of Congress Cataloging-in-Publication Data

Dedekind, Richard, 1831–1916.
    [Theorie der algebraischen Functionen einer Veränderlichen. English]
    Theory of algebraic functions of one variable / Richard Dedekind and Heinrich Weber ; Translated and introduced by John Stillwell.
        p. cm. — (History of mathematics ; v. 39)
    Includes bibliographical references and index.
    ISBN 978-0-8218-8330-3 (alk. paper)
    1. Algebraic functions.   2. Geometry, Algebraic. I. Weber, Heinrich, 1842–1913. II. Title.

QA341.D4313   2012
512.7′3—dc23
                                      2012011949

THEORY OF
ALGEBRAIC FUNCTIONS
OF ONE VARIABLE

Contents

Preface

Dedekind and Weber's 1882 paper on algebraic functions of one variable is one of the most important papers in the history of algebraic geometry. It changed the direction of the subject, and established its foundations, by introducing methods from algebraic number theory. Specifically, they used rings and ideals to give rigorous proofs of results previously obtained, in nonrigorous fashion, with the help of analysis and topology. Also, by importing ideas from number theory, the paper revealed the deep analogy between number fields and function fields—an analogy that continues to benefit both number theory and geometry today.

The influence of the paper is obvious in 20th-century algebraic geometry, where the role of arithmetic/algebraic methods has increased enormously in both scope and sophistication. But, as the sophistication of algebraic geometry has increased, so has its detachment from its origins. While the Dedekind-Weber paper continues to be cited, I venture to guess that few modern algebraic geometers are familiar with its contents. There are a few useful commentaries on the paper, but those that I know seem to focus on a few of the concepts used by Dedekind and Weber, while ignoring others. And, of course, fewer mathematicians today are able to read the language in which the paper was written (and I don't mean only the German language, but also the mathematical language of the 1880s).

I therefore believe that it is time for an English edition of the paper, with commentary to assist the modern reader. My commentary takes the form of a Translator's Introduction, which lays out the historical background to Dedekind and Weber's work, plus section-by-section comments and footnotes inserted in the translation itself. The comments attempt to guide the reader through the original text, which is somewhat terse and unmotivated, and the footnotes address specific details such as nonstandard terminology. The historical background is far richer than could be guessed from the Dedekind-Weber paper itself, including such things as Abel's results in integral calculus, Riemann's revolutionary approach to complex analysis and his discoveries in surface topology, and developments in number theory from Euler to Dedekind. The background is indeed richer than some readers may care to digest, but it is a background against which the clarity and simplicity of the Dedekind-Weber theory looks all the more impressive.

I hope that this edition will be of interest to several classes of readers: historians of mathematics who seek an annotated edition of one of the classics, mathematicians interested in history who would like to know where modern algebraic geometry came from, students of algebraic geometry who seek motivation for the concepts they are studying, and perhaps even algebraic geometers who have not had time to catch up with the origins of their discipline. (It seems to an outsider that just the modern literature on algebraic geometry would take more than a lifetime to absorb.)

This translation was originally written in the 1990s, but in 2011 I was motivated to revise it and write an introduction in order to prepare for a summer school presentation on ideal elements in mathematics. I have also compiled a bibliography and index. The bibliography is mainly for the Translator's Introduction, but it is occasionally referred to in the commentary on the translation, so I have placed it after the translation.

The summer school, PhilMath Intersem, was organized by Mic Detlefsen, and held in Paris and Nancy in June 2011. I thank Mic for inviting me and for support during the summer school. I also thank Monash University and the University of San Francisco for their support while I was researching this topic and writing it up. Anonymous reviewers from the AMS have been very helpful with some technical details of the translation, and I also thank Natalya Pluzhnikov for copyediting. Finally, I thank my colleague Tristan Needham, my wife Elaine, and son Robert for reading the manuscript and saving me from some embarrassing errors.

John Stillwell

South Melbourne, 1 May 2012

Translator's Introduction

1. Overview

Modern algebraic geometry has deservedly been considered for a long time as an exceedingly complex part of mathematics, drawing practically on every part to build up its concepts and methods and increasingly becoming an indispensable tool in many seemingly remote theories. It shares with number theory the distinction of having one of the longest and most intricate histories among all branches of our science, of having always attracted the efforts of the best mathematicians in each generation, and of still being one of the most active areas of research.

Dieudonné (1972), p. 827.

It seems to me that, in the spirit of the biogenetic law, the student who repeats in miniature the evolution of algebraic geometry will grasp the logic of the subject more clearly.

Shafarevich (1994), p. vii.

Richard Dedekind and Heinrich Weber first worked together in 1874, as co-editors of Riemann's collected works. Weber was called into this project as a replacement for Clebsch, who had died unexpectedly of diptheria, and his expertise in mathematical physics complemented Dedekind's expertise in pure algebra and analysis. The fruit of this collaboration was their joint paper, Dedekind and Weber (1882), a ground-breaking contribution to the understanding and advancement of Riemann's ideas. *Theorie der algebraischen Functionen der einer Veränderlichen* (theory of algebraic functions of one variable) revolutionized algebraic geometry by introducing methods of algebraic number theory into the subject. This made possible the first rigorous proofs of theorems discovered with the help of physical intuition, and opened the way to an extension of algebraic-geometric concepts from the complex numbers to arbitrary fields.

In a sense, the paper is a sequel to Dedekind (1877), a long paper in which Dedekind expounded his theory of ideals and their applications to number theory. However, Dedekind and Weber give a self-contained exposition of their theory, which is at some points simpler than the ideal theory for algebraic numbers.

Like Dedekind (1877), the Dedekind-Weber paper starts with the concept of field, but this time it is a field of *functions*, the "algebraic functions of one variable." Following the example of number theory, they distinguish the ring of *integers* of this field, then the *primes*, and finally the *ideals*. As in number theory, it turns out that ideals are crucial to complete the analogy with the traditional arithmetic of integers. However, in the context of algebraic functions, ideals prove to be important in other ways, and indeed a more general idea that they call "polygons" is needed.

To show the value of these new ideas, Dedekind and Weber gave new proofs of two great theorems: *Abel's theorem* of Abel (1841) and the *Riemann-Roch theorem* of Riemann (1857) and Roch (1864). These theorems are as timely today as they were in 1882, but they require some introduction, which Dedekind and Weber do not supply. I would therefore like to present some historical background to these theorems, and to the theory of algebraic functions itself, with copious examples. In some ways, this introduction is a sequel to my introduction to Dedekind (1877), though I will recapitulate some points to keep it self-contained.

In preparing this material I have been greatly assisted by the first 35 pages of Dieudonné (1985), an extraordinarily rich and insightful account of the development of algebraic geometry up to the Dedekind-Weber paper, and Koch (1991), which places this development against the general background of 19th-century mathematics. Another helpful overview is the chapter by Geyer in Dedekind et al. (1981). As will become apparent, much of the algebra in modern algebraic geometry arose from problems in classical analysis, particularly the integral calculus. The first such result was the fundamental theorem of algebra, originally motivated by the desire to factorize polynomials for the purpose of integrating rational functions.

2. From Calculus to Abel's Theory of Algebraic Curves

> What a discovery is Abel's generalization of Euler's integral! I have never seen such a thing! But how can it be that this discovery, which could be the most important made in the mathematics of this century, and which was communicated to your Academy two years ago, has escaped the attention of you and your colleagues?
>
> Jacobi (1829) letter to Legendre, 14 March 1829.

When calculus was developed in the 17th century, the first really hard problems were problems of integration. This was especially true of the Leibniz approach, which sought integrals in "closed form," that is, in terms of functions from the small class known as "elementary." These are the algebraic functions, together with functions arising from them by composition with the exponential function and its relatives, the logarithm, circular functions, and inverse circular functions.

The only broad class of functions that can be integrated in Leibniz's sense are the rational functions, that is, the functions of the form $r(x) = p(x)/q(x)$, where p and q are polynomials. Any rational function can be integrated because the denominator $q(x)$ may be split into linear factors $(x - a)$, by the fundamental theorem of algebra, and the quotient $p(x)/q(x)$ may then be decomposed into partial fractions of the form $x^m/(x - a)^n$, which have rational integrals in all cases except

$$\int \frac{dx}{x - a} = \log(x - a) + \text{constant.}$$

Thus the integral of a rational function is itself a rational function, with the possible exception of some terms of the form $\log(x - a)$.

(In elementary calculus courses this simple picture is confused by the presence of partial fractions such as $1/(x^2 + 1)$, the integral of which is usually taken to be $\tan^{-1} x + \text{constant}$. However, we have

$$\frac{1}{x^2 + 1} = \frac{1/2i}{x - i} - \frac{1/2i}{x + i},$$

so we can also express $\int dx/(x^2+1)$ as a sum of logarithms, namely

$$\int \frac{dx}{x^2+1} = \frac{1}{2i}\int \frac{dx}{x-i} - \frac{1}{2i}\int \frac{dx}{x+i} = \frac{1}{2i}\log(x-i) - \frac{1}{2i}\log(x+i).$$

This was first done, albeit with some confusion about the meaning of complex logarithms, by Johann Bernoulli (1702). Around 1800, when the fundamental theorem of algebra was finally proved, the meaning of complex numbers became better understood, and it became increasingly clear that they play an important role in the theory of integrals.)

When the rational functions are extended by as little as the square root function, the resulting integrals quickly fall outside the class of elementary functions. A famous example is the *lemniscatic* integral

$$\mathrm{sl}^{-1}(x) = \int_0^x \frac{dt}{\sqrt{1-t^4}},$$

so-called because it expresses the arc length of the lemniscate of Jakob Bernoulli (1694), shown in Figure 1.

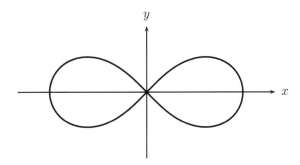

FIGURE 1. The lemniscate of Bernoulli

This curve has cartesian equation $(x^2+y^2)^2 = x^2-y^2$, and its arc length cannot be expressed in terms of elementary functions of x and y. However, Fagnano (1718) discovered that the lemniscatic integral satisfies an arc-length *doubling formula*

$$2\int_0^x \frac{dt}{\sqrt{1-t^4}} = \int_0^{2x\sqrt{1-x^4}/(1+x^4)} \frac{dt}{\sqrt{1-t^4}},$$

and Euler (1768) generalized Fagnano's formula to an arc-length *addition formula*

$$(*)\qquad \int_0^x \frac{dt}{\sqrt{1-t^4}} + \int_0^y \frac{dt}{\sqrt{1-t^4}} = \int_0^{(x\sqrt{1-y^4}+y\sqrt{1-x^4})/(1+x^2y^2)} \frac{dt}{\sqrt{1-t^4}}.$$

These results are analogous to properties of the *inverse sine integral*

$$\theta = \sin^{-1}x = \int_0^x \frac{dt}{\sqrt{1-t^2}},$$

which are derivable from basic properties of the sine and cosine functions. For example, the familiar angle-doubling formula

$$\sin 2\theta = 2\sin\theta\cos\theta = 2\sin\theta\sqrt{1-\sin^2\theta},$$

implies that

$$2\theta = \sin^{-1}(2\sin\theta\sqrt{1-\sin^2\theta}),$$

which gives the angle-doubling formula for integrals:

$$2\sin^{-1}x = 2\int_0^x \frac{dt}{\sqrt{1-t^2}} = \int_0^{2x\sqrt{1-x^2}} \frac{dt}{\sqrt{1-t^2}}.$$

And the familiar angle addition formula,

$$\sin(\theta + \varphi) = \sin\theta\cos\varphi + \cos\theta\sin\varphi$$

$$= \sin\theta\sqrt{1-\sin^2\varphi} + \sin\varphi\sqrt{1-\sin^2\theta},$$

implies

$$\theta + \varphi = \sin^{-1}\left(\sin\theta\sqrt{1-\sin^2\varphi} + \sin\varphi\sqrt{1-\sin^2\theta}\right),$$

which gives the addition formula for arcsine integrals:

$$\int_0^x \frac{dt}{\sqrt{1-t^2}} + \int_0^y \frac{dt}{\sqrt{1-t^2}} = \int_0^{x\sqrt{1-y^2}+y\sqrt{1-x^2}} \frac{dt}{\sqrt{1-t^2}}.$$

Thus in both cases we find that a sum of two integrals, $\int_0^x f(t)\,dt + \int_0^y f(t)\,dt$, can be simplified to a single integral, $\int_0^z f(t)\,dt$, where z is an algebraic function of x and y.

It so happens that the integrand $1/\sqrt{1-t^2}$ of the inverse sine integral can be *rationalized* by the change of variable $t = 2s/(1+s^2)$—not so surprisingly, since the inverse sine is an elementary function—so we can eliminate the integral altogether in this case. However, in the case of the lemniscatic integral, reducing the sum of two integrals to one is the best we can do. The integrand $1/\sqrt{1-t^4}$ *cannot* be rationalized by a change of variable, and indeed Jakob Bernoulli (1704) made a remarkable attempt to prove this, using the theorem of Fermat that the equation $X^4 - Y^4 = Z^2$ has no solution in positive integers X, Y, Z. His attempt fell short, because it is not enough to know this theorem for integers X, Y, Z. But it is enough to know it for *polynomials* $X(t), Y(t), Z(t)$, and indeed polynomials behave enough like integers that Fermat's proof can be replayed for polynomials, though no one noticed this in Bernoulli's time.

Thus there is an essential difference between the ordinary sine function and the lemniscatic sine function, sl, defined as the inverse of the lemniscatic integral. Nevertheless there are enough similarities to enable the development of a theory of the lemniscatic sine function. This was begun by Gauss in 1796, and extended to a general theory of the so-called *elliptic functions* by Abel and Jacobi in the 1820s. Like the circular functions, the elliptic functions satisfy addition formulas and they are *periodic*, only more so. Just as the sine and cosine have *period* 2π, in the sense that

$$\sin(\theta + 2\pi) = \sin\theta, \quad \cos(\theta + 2\pi) = \cos\theta,$$

an elliptic function f has *two* periods ω_1, ω_2, in the sense that

$$f(z + \omega_1) = f(z), \quad f(z + \omega_2) = f(z).$$

The periods ω_1, ω_2 are complex numbers whose ratio is not real. For example, Gauss discovered that the two periods of sl are ϖ and $i\varpi$, where

$$\varpi = 2 \int_0^1 \frac{dt}{\sqrt{1 - t^4}}.$$

The double periodicity of elliptic functions was first explained by algebraic manipulation of integrals, but Riemann (1851) found a far more transparent geometric explanation (not unlike explaining the period 2π of sine and cosine by referring to the circle), which we will come to later.

The theory of elliptic functions was the first great advance in integral calculus since the integration of rational functions. Nevertheless, this theory only scratched the surface of a huge and important world of calculus: the integrals of *algebraic functions*; that is, integrals of the form

$$\int g(s, t)\, dt, \quad \text{where } s \text{ satisfies a polynomial equation} \quad P(s, t) = 0.$$

The lemniscatic integral is $\int dt/s$, where $s^2 = 1 - t^4$, and the general theory of elliptic functions deals with the integrals $\int dt/s$ where s^2 equals a polynomial of degree 3 or 4 in t. But what can one say, for example, about the integral

$$\int_0^x \frac{dt}{\sqrt{1 - t^6}} \ ?$$

It turns out that this integral does *not* satisfy an addition formula

$$\int_0^x \frac{dt}{\sqrt{1 - t^6}} + \int_0^y \frac{dt}{\sqrt{1 - t^6}} = \int_0^z \frac{dt}{\sqrt{1 - t^6}},$$

where z is an algebraic function of x and y. However, Abel (1841) discovered a wonderful substitute for an addition formula: *any sum of integrals,*

$$\int_0^{x_1} \frac{dt}{\sqrt{1 - t^6}} + \cdots + \int_0^{x_m} \frac{dt}{\sqrt{1 - t^6}}$$

is equal to the sum of two *integrals*

$$\int_0^{z_1} \frac{dt}{\sqrt{1 - t^6}} + \int_0^{z_2} \frac{dt}{\sqrt{1 - t^6}}, \quad \text{where } z_1, z_2 \text{ are algebraic functions of } x_1, \ldots, x_m,$$

plus some "trivial" algebraic and logarithmic terms.

This result is only an illustration of the amazingly general:

Abel's Theorem. *For any integral of the form $\int g(s, t)\, dt$, where g is a rational function and s and t are connected by a polynomial relation $P(s, t) = 0$, there is a number p such that any sum of integrals*

$$\int_0^{x_1} g(s, t)\, dt + \cdots + \int_0^{x_m} g(s, t)\, dt$$

equals a sum of at most p integrals

$$\int_0^{z_1} g(s, t)\, dt + \cdots + \int_0^{z_p} g(s, t)\, dt,$$

where $z_1, \ldots, z_p$ are algebraic functions of $x_1, \ldots, x_m$, plus terms that are either rational functions or their logarithms.

The number p depends only on the polynomial P. It was later called the *genus* of the curve defined by $P(s, t) = 0$, and it too found a natural geometric interpretation in Riemann (1851), as we will see in the next section. In particular, the curve $s^2 = 1 - t^4$ that yields the lemniscatic integral has genus 1, because any sum of lemniscatic integrals reduces to one integral by repeated application of the addition formula (*). More generally, any *elliptic curve*[1] $s^2 = q(t)$, where $q(t)$ is of degree 3 or 4 without repeated roots, is of genus 1, because there is an addition formula for the corresponding integral $\int dt/s$.

Finally, any curve $s = P(w)$, $t = Q(w)$ parameterized by rational functions P and Q is of genus zero, because the corresponding integral $\int g(s,t)\,dt$ is the integral of the rational function $g(P(w), Q(w))Q'(w)$.

Abel submitted his paper to Cauchy in 1826 but, due to inattention by the mathematicians of the Paris Academy, it was not published at the time. It was noticed by Jacobi, however, who in 1829 wrote the letter to Legendre quoted at the beginning of this section. Even the intervention of Jacobi failed to wake up the Academicians, and Abel's paper did not appear until 1841, long after Abel had died. There is a further excruciating twist to this story of neglected genius. The other mathematician notoriously ignored by the Paris Academy, Evariste Galois, also seems to have discovered Abel's theorem, independently of Abel, but some years later. It is mentioned in his letter to Auguste Chevalier, Galois (1846), written on the night before his death in 1832. He states the theorem without proof, but with some additional remarks that suggest that he already had some of the ideas developed by Riemann 20 years later.

3. Riemann's Theory of Algebraic Curves

> It is quite a paradox that in the work of this prodigious genius, out of which algebraic geometry emerges entirely regenerated, there is almost no mention of algebraic curve; it is from his theory of algebraic *functions* and their integrals that all of the birational geometry of the nineteenth and the beginning of the twentieth century issues.
>
> Dieudonné (1985), p. 18.

In the 1850s, two papers by Bernhard Riemann[2] completely changed the face of complex analysis and algebraic geometry. Riemann (1851) and Riemann (1857) viewed algebraic curves in a new way, as what we now call *Riemann surfaces*. In retrospect, this development seems unsurprising and even inevitable. Since around 1800, mathematicians had become used to the idea that the complex "line" $\mathbb{C}$ was geometrically a plane, so the idea that a complex "curve" should be some kind of surface was just over the horizon. Nevertheless, Riemann's description of these surfaces was received skeptically by most of his contemporaries. The underlying topological ideas, though very intuitive and persuasive, did not yet have a rigorous foundation. And, to make matters worse, Riemann made connections between topology and analysis by appealing to physics. Then, as now, this was considered mathematically dubious.

---

[1] The name "elliptic" became attached to the curves of genus 1 because the corresponding integrals ("elliptic integrals") include the integral for the arc length of the ellipse. Unfortunately, the ellipse itself has genus 0, and hence is *not* an elliptic curve.

[2] Page numbers in references to these papers in this Introduction refer to the original papers. However, many readers will find it more convenient to consult the English translation of Riemann's works, Riemann (2004). To make this easier to do, I also give section numbers, which are the same in the original papers and in the translation.

But if Riemann's proofs were not rigorous, his results were so stunning that they demanded explanation, and this became the task of later mathematicians, among them Dedekind and Weber.

Today, the necessary foundations of topology and analysis have been constructed, so we have the luxury of describing Riemann's ideas in informal terms similar to his own. I think that it is useful to do so, because some of the algebraic concepts devised by Dedekind and Weber are scarcely comprehensible if one has not seen the topological concepts they replace. In particular, I doubt that readers should be confronted with the "ramification ideal" before they have seen a picture of "ramification," or "branching." Such a picture was given in Neumann (1865), the first textbook on Riemann's theory (Figure 2).

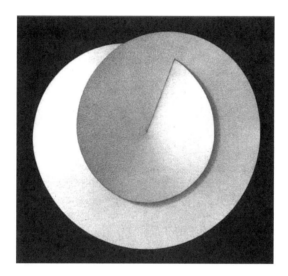

FIGURE 2. Neumann's picture of a branch point

This picture springs to mind when one attempts to visualize the curve $y^2 = x$ for *complex* variables x and y or, equivalently, the "two-valued function" $y = \pm\sqrt{x}$. Riemann imagined the two values $+\sqrt{x}$ and $-\sqrt{x}$ lying above x on a *two-sheeted covering* of the plane $\mathbb{C}$, as shown in Figure 3. Notice that, as x moves continuously once around a circle, the corresponding point $\sqrt{x}$ moves continuously around the lower sheet, then the upper sheet, of the two-sheeted covering, eventually taking the value $-\sqrt{x}$ that also lies above x. Thus the function $\sqrt{x}$ becomes "single-valued" on the covering surface.

The point $x = 0$ at which the two sheets fuse is called a *branch point* or *ramification point* of the covering, because one used to speak of the "branches of the multi-valued function"—in this case the two "branches" are $\sqrt{x}$ and $-\sqrt{x}$. The awkward feature of the picture—that the two sheets appear to pass through each other—is a result of representing the relation $y = x^2$ in three dimensions, one fewer than the four dimensions it really requires. One can visually add a fourth dimension, "shade of gray," to the sheets to avoid their meeting in the fourth dimension. This has actually happened in the Neumann picture, where one sheet is white where they appear to meet and the other is dark gray.

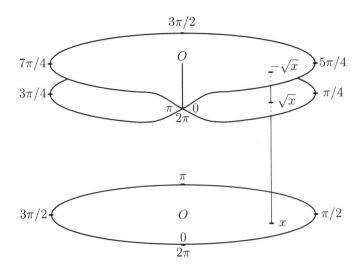

FIGURE 3. Branch point for the square root

Just as the curve $y^2 = x$ has a branch point of the two sheets at $x = 0$, the curve $y^n = x$ has a branch point of n sheets. An arbitrary algebraic curve $P(x, y) = 0$, where P is a polynomial of degree n, is an n-sheeted covering of $\mathbb{C}$ with a finite number of branch points. Since behavior of a curve at infinity is important, Riemann (1857) (Section 1, p. 117) extended $\mathbb{C}$ by a point ∞, and the resulting set $\mathbb{C} \cup \{\infty\}$ can be viewed as a *sphere* via the stereographic projection map shown in Figure 4. This idea is made explicit in Neumann (1865), p. 132. Under stereographic projection, each point $z \in \mathbb{C}$ corresponds to a point z' on the sphere other than the north pole N, and N itself naturally corresponds to ∞.

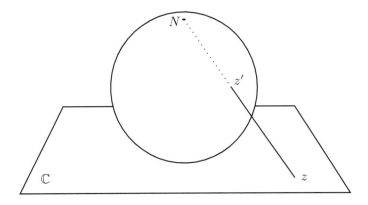

FIGURE 4. Stereographic projection of the sphere to $\mathbb{C} \cup \{\infty\}$

Corresponding to this completion of $\mathbb{C}$ to a sphere, we have a completion of each algebraic curve to a finite-sheeted covering of the sphere with finitely many branch points. Riemann realized that the covering surface S is topologically characterized by none other than Abel's number p, later dubbed the *genus* (or *Geschlecht* in German) by Clebsch (1865). Riemann described p topologically as half the number of closed cuts needed to make S simply connected (that is, such that any closed curve can be contracted to a point). In this case the resulting simply connected surface is a polygon with $4p$ sides. Möbius (1863) gave an even simpler interpretation of p, by showing that each Riemann surface is topologically equivalent to a member of the sequence of surfaces shown in Figure 5, namely, the one with p "holes."

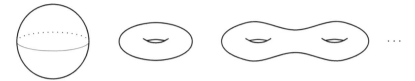

FIGURE 5. Riemann surfaces of genus 0, 1, 2, ...

As an example, consider the elliptic curve

$$y^2 = x(x-1)(x+1).$$

This curve is a two-sheeted cover of the sphere, with branch points like that shown in Figure 3 at $x = 0, 1, -1, \infty$. If we slit the sheets by cuts from 0 to ∞, and from 1 to -1, then, in order to obtain the branching, the edges of the cuts need to be identified so that the like-labeled edges shown in Figure 6 come together.

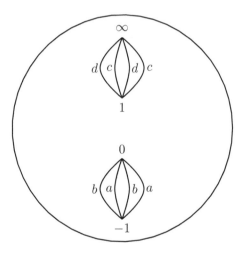

FIGURE 6. How edges are identified at branch points

But we can make a surface that is topologically the same by separating the two sheets before making the identifications, as shown in Figure 7.

The resulting surface is topologically a torus, shown in more familiar form in Figure 8. Thus Abel's number $p = 1$ agrees with the topological genus of the torus, because the torus has one "hole." (Notice that if 0 and ∞ are the only branch points, as is the case with $y^2 = x$, then the result of joining the two sheets is topologically a sphere, so the genus of $y^2 = x$ is zero.)

Moreover, as promised in the previous section, we can now see the reason for the two periods of elliptic functions associated with the curve $y^2 = x(x-1)(x+1)$. The periods are integrals over independent closed paths on the torus surface, such as the paths C_1 and C_2 shown in Figure 8.

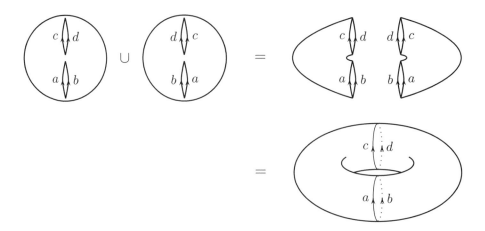

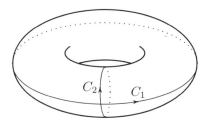

FIGURE 7. Identifying edges after separating the sheets

FIGURE 8. Independent closed paths on the torus

4. The Riemann-Hurwitz Formula

For a simply connected surface, spread over a finite region of the z-plane, there is a relationship between the number of simple branch points and the number of windings of its boundary curve .... From this there results a relation, for a multiply connected surface, between these numbers and the number of transverse cuts needed to transform it into a simply connected surface. This relation, which does not depend on metric considerations and belongs to *analysis situs*, can be derived as follows ...

Riemann (1857), Section 7, pp. 127–128.

As we saw in the previous section, Riemann viewed an algebraic curve as a surface covering the sphere $\mathbb{C} \cup \{\infty\}$. For a curve of degree n the covering is of n "sheets": that is, above each point of $\mathbb{C} \cup \{\infty\}$ there are n different points of the curve, with the exception of finitely many *ramification points* where two or more sheets come together. Riemann (1857) showed that the genus can be computed from the degree n and numbers e_P that give the number of sheets fused together at a ramification point P. Because of a generalization made by Hurwitz (1891), the method is now called the *Riemann-Hurwitz formula*.

The Riemann-Hurwitz formula is most easily explained in terms of the oldest manifestation of the genus concept, the *Euler characteristic*, which goes back to Euler (1752). Euler observed that, for any decomposition of the sphere into F faces, E edges, and V vertices one has $V - E + F = 2$. More generally, the number $V - E + F$ is invariant for any surface, and it equals $2 - 2p$, where p is the genus. The invariance of $V - E + F$ follows by observing that any two partitions of a surface have a common subdivision, obtained by superimposing one on the other. (This argument is merely plausible, since an edge in one partition may intersect edges of the other partition infinitely often, but Riemann used it in Riemann (1851), Section 6, p. 10.) And any subdivision may be obtained by a series of elementary subdivisions of the following two types:

(1) Subdividing an edge by a new vertex.
(2) Subdividing a face by a new edge connecting two of its vertices.

The first increases both V and E by 1, the second increases both E and F by 1, so neither changes $V - E + F$.

Thus $V - E + F$ is invariant and, by computing it for a standard subdivision of the genus p surface (such as the one shown for the torus in Figure 8), one finds that $V - E + F = 2 - 2p$. This quantity is now called the Euler characteristic, $\chi(\mathcal{S})$, of the surface $\mathcal{S}$ of genus p. It enables us to compute the genus from a decomposition of the surface into vertices, edges, and faces, which is particularly convenient when $\mathcal{S}$ is realized as a ramified covering of the sphere. If $\mathcal{S}$ were simply an n-sheeted covering of the sphere then we should have $\chi(\mathcal{S}) = 2n$, since 2 is the Euler characteristic of the sphere and $\mathcal{S}$ has n vertices over each vertex on the sphere, n edges over each edge, and n faces over each face.

However, when $n > 1$, the covering has ramification points, so we have to adjust the count of vertices. We assume that the ramification points are included among the vertices on the sphere, in which case the count of vertices on $\mathcal{S}$ must be reduced by $e_P - 1$ at each ramification point P. On the other hand, the numbers of edges and faces on $\mathcal{S}$ are still n times the corresponding numbers on the sphere, so we have

$$\chi(\mathcal{S}) = 2n - \sum_P (e_P - 1).$$

Thus the genus p of $\mathcal{S}$ depends only on the degree n of the curve and its ramification numbers e_P.

As an example, consider the 2-sheeted covering of the sphere with the four branch points $P = -1, 0, 1, \infty$ in the previous section. At each of these points $e_P = 2$, so the Riemann-Hurwitz formula gives $2 - 2p = 2 \cdot 2 - 4 = 0$. Hence $p = 1$, as obtained previously.

Since $\chi(\mathcal{S}) = V - E + F = 2 - 2p$, the above formula for $\chi(\mathcal{S})$ may be written as the following formula for genus:

$$p = \frac{1}{2}w - n + 1,$$

where $w = \sum_P (e_P - 1)$ is the sum of the ramification numbers. This formula is used as the *definition* of genus in §24 of the Dedekind-Weber paper.

The determination of genus by degree and ramification numbers is crucial to Dedekind and Weber's program of deriving Riemann's results by algebraic methods. Their algebraic definition of a "Riemann surface" $\mathcal{S}$ lacks geometric structure, so it is not meaningful to decompose $\mathcal{S}$ by vertices, edges, and faces (nor can one make "cuts" in it between the ramification points, as in Figure 6). However, it *is* possible to show that there are n distinct points of S over each point of the sphere $\mathbb{C} \cup \{\infty\}$, with the exception of finitely many ramification points P, and the ramification number e_P at each such point can be defined algebraically. Thus they can define genus in a way that agrees with Riemann's concept of genus, by means of the Riemann-Hurwitz formula.

5. Functions on Riemann Surfaces

A system of like-branching algebraic functions and their integrals is the main object of our study; but instead of proceeding from expressions for these functions we define them by their discontinuities with the help of Dirichlet's principle.

Riemann (1857), Section 1, p. 117.

It certainly is somewhat daring to infer the existence of U from its hydrodynamic significance.

Weyl (1964), p. 107.

With the insight gained from his view of algebraic curves as surfaces covering the plane $\mathbb{C}$ or the sphere $\mathbb{C} \cup \{\infty\}$, Riemann was able to generalize Cauchy's theory of integration on $\mathbb{C}$ to integration on arbitrary algebraic curves. To see where this may lead, let us first recall some of Cauchy's main results, and some of their immediate consequences. Cauchy developed his theory, in stages of increasing generality, between 1814 and 1831. For a very readable and insightful account of this period, see Smithies (1997).

The fundamental result is *Cauchy's theorem*, according to which

$$\int_C f(z)\,dz = 0,$$

where f is differentiable or *holomorphic* (and hence not infinite) on and inside the closed curve C in the complex plane $\mathbb{C}$. It follows that $\int_a^b f(z)\,dz$ does not depend on the path chosen between a and b, as long as the paths lie in a simply connected region of $\mathbb{C}$ where f does not become infinite.

Of course, it is a different story for functions that *do* become infinite, such as $f(z) = 1/z$ at $z = 0$. We find, for example, that

$$\int_C \frac{dz}{z} = 2\pi i,$$

where C is a clockwise path around the unit circle. Nevertheless, the value of the integral of $1/z$ around a closed path C depends only slightly on C—it is the same around any path C' to which C can be deformed without crossing the point $z = 0$.

This example is generalized in *Cauchy's integral formula* and *Cauchy's residue theorem*. The integral formula says that

$$f(a) = \frac{1}{2\pi i} \int_C \frac{f(z)\,dz}{z-a},$$

where C is a circle with center a in a region where f is differentiable, and it has the consequence that

$$f^{(n)}(a) = \frac{n!}{2\pi i} \int_C \frac{f(z)\,dz}{(z-a)^{n+1}},$$

so a differentiable complex function f in fact has derivatives of all orders. It also follows that f has a Taylor series expansion in the neighborhood of a point where f' exists, and from this we can conclude that if $f(a_k) = 0$ for a convergent sequence of points a_k then f is constantly zero. Another important consequence is the so-called *Liouville's theorem*: if f is differentiable and *bounded* on $\mathbb{C}$ then f is constant. (This theorem was discovered by Liouville in 1844 but first published by Cauchy (1844); see Lützen (1990), p. 124.)

The residue theorem generalizes the property of the function $1/z$ to an arbitrary *meromorphic* function f—one that is differentiable except at isolated *poles* $z = a$, in the neighborhood of which

$$f(z) = c_{-h}(z - a)^{-h} + \cdots + c_{-1}(z - a)^{-1} + c_0 + c_1(z - a) + c_2(z - a)^2 + \cdots .$$

The integer h is called the *multiplicity* or *order* of the pole $z = a$, and the coefficient c_{-1} is called the *residue* of f at $z = a$. For such a function,

$$\int_C f(z)\, dz = 2\pi i c_{-1},$$

where C is a clockwise path around $z = a$, small enough not to touch or enclose any other pole of f. Thus the value of the integral depends only on the residue. More generally, if C is a path running clockwise around poles of f at which the residues are $r_1, \ldots, r_m$, then

$$\int_C f(z)\, dz = 2\pi i(r_1 + \cdots + r_m).$$

With these theorems we can find *all the functions f on $\mathbb{C} \cup \{\infty\}$ that are meromorphic. They are precisely the rational functions.* A rational function

$$f(z) = \frac{p(z)}{q(z)}, \quad \text{where } p \text{ and } q \text{ are polynomials},$$

is differentiable except at its finitely many poles where $q(z) = 0$, so a rational function is certainly meromorphic.

Conversely, if $f(z)$ is meromorphic on $\mathbb{C} \cup \{\infty\}$ and f is not the constant zero, then it can have only finitely many zeros and poles. (If not, there are infinitely many zeros or poles in a neighborhood of ∞. Infinitely many zeros force f to be the constant zero, and infinitely many poles destroy differentiability near infinity.) Now, multiplying $f(z)$ by a rational function $g(z)$ that cancels its zeros and poles, we obtain a function $h(z) = f(z)g(z)$ that is finite and nonzero everywhere except possibly at $z = \infty$. In any case, either $h(z)$ or $1/h(z)$ is bounded, hence constant by Liouville's theorem. Thus $f(z)$ is a constant multiple of the rational function $1/g(z)$, hence rational itself.

Now a rational function $f(z) = p(z)/q(z)$ is determined, up to a constant multiple, by its zeros and poles and their orders. Assuming any common factors of $p(z)$ and $q(z)$ have been canceled, the zeros of $p(z)$ give zeros of $f(z)$ and the zeros of $q(z)$ give poles of $f(z)$. In addition, if $\deg(p) \neq \deg(q)$, there will be a zero or pole at ∞ according as $\deg(p) < \deg(q)$ or $\deg(p) > \deg(q)$, making the total order of zeros equal to the total order of poles.

To summarize: *a meromorphic function on $\mathbb{C} \cup \{\infty\}$ is determined, up to a constant multiple, by a finite set of zeros and poles, with associated orders. A function with given zeros and poles actually exists (namely, a rational function) provided that the total order of zeros equals the total order of poles.*

Thus Riemann's program of "defining a function by its discontinuities" is easily carried out on $\mathbb{C} \cup \{\infty\}$, a Riemann surface of genus zero. We need only understand the "discontinuities" to be the zeros and poles. Implicitly, Cauchy had the result, though perhaps not the means of expressing it in terms of functions on a surface. However, extending this result to *all* surfaces of genus zero is already a significant problem. It was solved by Riemann (1851) via the *Riemann mapping theorem*, a difficult result that he proved by appealing to what he called the *Dirichlet principle*. The Dirichlet principle is a powerful method for proving the existence of functions with given properties (such as specified zeros and poles) on surfaces, but it stems from physical intuition about the flow of electricity and was not proved in a form suitable for Riemann's applications until after 1900.

It should be stressed that this remarkably simple view of meromorphic functions on $\mathbb{C} \cup \{\infty\}$ is made possible by the point ∞. We would now say that the role of ∞ is to make the surface *compact*. Compactness ensures that any infinite set of points has a limit, which makes possible the above argument that a meromorphic function on $\mathbb{C} \cup \{\infty\}$ is rational. The meromorphic functions on $\mathbb{C}$ are a much less manageable class, since they include functions that are not even algebraic, such as e^z. The Riemann (1857) view of an algebraic curve as a surface covering the sphere led him to an equally simple view of the meromorphic functions on such a curve—one that would be the starting point of the Dedekind-Weber theory—they are *algebraic functions*, of degree bounded by the degree of the curve.

To see why, suppose that f is s meromorphic function on a Riemann surface $\mathcal{S}$, so f has only finitely many poles on $\mathcal{S}$. If $\mathcal{S}$ is of degree n then there are n points $x_1, \ldots, x_n$ (not necessarily distinct) over each $x \in \mathbb{C} \cup \{\infty\}$. Consider the values $f(x_1), \ldots, f(x_n)$ of f at these points. Then the "multi-valued function" $y = f(x)$ satisfies

$$(y - f(x_1)) \cdots (y - f(x_n)) = 0.$$

Expanding the left side, we get

(*) $$y^n + a_{n-1} y^{n-1} + \cdots + a_1 y + a_0 = 0,$$

where

$$a_0 = (-1)^n f(x_1) \cdots f(x_n),$$

$$\vdots$$

$$a_{n-1} = -(f(x_1) + \cdots + f(x_n))$$

are the elementary symmetric functions of $f(x_1), \ldots, f(x_n)$. Because of their symmetry, $a_0, \ldots a_{n-1}$ are well-defined meromorphic functions of $x \in \mathbb{C} \cup \{\infty\}$. So, $y = f(x)$ satisfies a polynomial equation (*) of degree n whose coefficients are rational functions of x, and this means that $f(x)$ is an algebraic function of x, of degree at most n. This is essentially the argument of Riemann (1857), Section 5, p. 123.

On a surface of genus $p > 0$ it could likewise be proved that a meromorphic function is determined, up to a constant multiple, by its zeros and poles, and that the total order of zeros equals the total order of poles. However, it could also be proved that there are further *constraints on existence*, arising from closed paths that do not bound a piece of the surface and the corresponding periods of integrals. For example, on a surface of genus 1 there is no function with one pole of order 1 and one zero of order 1. This raises the problem of finding functions, with given zeros and poles, that satisfy the known constraints.

For genus $p = 1$, constraints were found by Abel (1827), and he was also able to prove that any finite set of zeros and poles satisfying these constraints could be realized by a meromorphic function, that is, by an elliptic function. He did this with the help of "expressions for these functions," which he and Jacobi had already raised to a fine art. "Expressions" for functions on surfaces of higher genus were much less developed in Riemann's time, so he actually made a virtue of a necessity by appealing to Dirichlet's principle in order to prove their existence. The trouble was, Weierstrass (1870) showed that the Dirichlet principle *fails* in certain cases, so Riemann's methods were under suspicion for some decades.

Riemann (1857) considered functions on a surface of genus p with simple poles (that is, poles of order 1) at r given points. These functions form a vector space (over the field $\mathbb{C}$) whose dimension l Riemann proved to satisfy

$$l \geq r - p + 1$$

(Riemann's inequality). Riemann did not use the language of vector spaces and dimensions,[3] which did not yet exist; he said that the functions have l "arbitrary constants." The inequality generalizes to the case where the poles have multiplicities $d_1, \ldots, d_r$, in which case

$$l \geq (d_1 + \cdots + d_r) - p + 1.$$

Riemann's student Roch (1865) turned the inequality into an equality by interpreting the difference

$$l - [(d_1 + \cdots + d_r) - p + 1]$$

as the dimension of a space of certain functions, today called the *canonical class*.

In the special case of meromorphic functions on $\mathbb{C} \cup \{\infty\}$ we can see that

$$l = (d_1 + \cdots + d_r) - p + 1,$$

because in this case $p = 0$ and the meromorphic functions with r poles are of the form

$$f(z) = \frac{p(z)}{k(z - p_1)^{d_1} \cdots (z - p_r)^{d_r}},$$

if ∞ is not one of the poles. In this case $p(z)$ can be any polynomial of degree at most $d_1 + \cdots + d_r$ (the degree of $q(z)$), and the space of such functions indeed has dimension $(d_1 + \cdots + d_r) + 1$, because there are $(d_1 + \cdots + d_r) + 1$ arbitrary constants in the definition of a polynomial of degree $(d_1 + \cdots + d_r)$. If ∞ *is* a pole, then its order is the difference $\deg(p) - \deg(q)$, and it is easily checked that $l = (d_1 + \cdots + d_r) + 1$ in this case also.

We will explain in Section 9 how Dedekind and Weber overhauled these ideas so as to avoid assuming the Dirichlet principle, and thereby transformed Riemann-Roch into a theorem of algebra. There were attempts to prove the Riemann-Roch theorem without appealing to analysis and topology before Dedekind and Weber, but these proofs were not completely general, as Dedekind and Weber indicate in the first few sentences of their paper. They point out that these previous attempts were

---

[3]Indeed, the concept is still struggling to emerge in Dedekind and Weber's paper. They use the term *Schaar* for what we call a vector space, but they reprove the basic vector space properties for each new *Schaar* that comes up. Dedekind formally described the vector space properties of a *Schaar* for the first time on pp. 467–468 of Dedekind (1894).

made "under certain restrictive assumptions about the singularities of the functions under consideration." This was the case, in particular, for the proof in Brill and Noether (1874), where the term "Riemann-Roch theorem" is first employed. For more details see Gray (1998). In any case, before discussing Dedekind and Weber, we should say a few words on the later development of Riemann's ideas, and their eventual vindication.

6. Later Development of Analysis on Riemann Surfaces

> Having reached Coutances, we entered an omnibus to go some place or other. At the moment when I put my foot on the step the idea came to me, without anything in my former thoughts seeming to have paved the way for it, that the transformations I had used to define the Fuchsian functions were identical with those of non-Euclidean geometry.
>
> From Poincaré's essay "Mathematical creation" in Poincaré (1918).

As explained in the previous section, Riemann viewed algebraic curves as surfaces, so that "multi-valued functions" such as $\sqrt{z}$ became single-valued and the genus p had a simple topological meaning. He also interpreted meromorphic functions in terms of *flows of electricity*, brought about by applying the poles of a battery to points of a surface covered by an infinitely thin layer of conducting material. (In fact, this is apparently where the word "pole" comes from.) An account of Riemann's theory in these frankly unrigorous, but intuitively helpful, terms was given by Klein (1882).

Klein's book (appearing in the same year as the Dedekind and Weber paper) made no advance towards a proof of Riemann's theorems, since the crucial Dirichlet principle was still assumed without proof, and the concept of surface remained vague. However, it did inspire other mathematicians to rigorize the questionable parts of Riemann's work. Hilbert (1904) proved a "Dirichlet principle" strong enough for Riemann's needs, and Weyl (1913) completed the theory with a precise definition of Riemann surfaces. In fact, Weyl's definition was soon imitated in all parts of complex analysis and differential geometry, where it was useful to have a concept of a *manifold with differentiable structure*, and where one wants to decide which structures are isomorphic.

This was not the only direction in which the concept of Riemann surface developed. Another was in *generalizing the theory of elliptic functions to higher genus*. To explain what there is to generalize, I will outline the basic facts about elliptic functions and their relation to Riemann surfaces.

Thanks to the work of Abel and Jacobi, elliptic functions were already well known in the 1820s as doubly-periodic functions on $\mathbb{C}$. As we have already mentioned in Section 3, the Riemann surface concept explains the two periods as integrals around certain closed curves on the torus, such as C_1 and C_2 in Figure 8. However, it is also possible to exhibit double periodicity in a *formula* for an elliptic function. The simplest possible formula is one due to Eisenstein (1847):

$$f(z) = \sum_{m,n \in \mathbb{Z}} \frac{1}{(z + m\omega_1 + n\omega_2)^2}.$$

Assuming that this series is meaningful (which it is, if the summation is interpreted properly), then it clearly remains the same if z is replaced by $z + \omega_1$ or $z + \omega_2$.

Thus f has two periods, ω_1 and ω_2. Almost as simple, and more standard, is the *pe-function* of Weierstrass (1863):

$$\wp(z) = \frac{1}{z^2} + \sum_{(m,n)\neq(0,0)} \left(\frac{1}{(z+m\omega_1+n\omega_2)^2} - \frac{1}{(m\omega_1+n\omega_2)^2} \right).$$

The Weierstrass function $\wp(z)$ is not so obviously periodic (one first proves this for $\wp'$), but it is easier to work with because its series is uniformly convergent except at the double poles where $z = m\omega_1 + n\omega_2$. For example, Weierstrass was able to show, by simple series manipulations, that

$$\wp'(z)^2 = 4\wp(z)^3 - g_2\wp(z) - g_3,$$

where g_2 and g_3 are certain constants depending on ω_1 and ω_2. Thus the functions $x = \wp(z)$ and $y = \wp'(z)$ parameterize the curve

$$y^2 = 4x^3 - g_2 x - g_3,$$

to which any curve of genus 1 happens to be birationally equivalent (for suitable choice of g_2 and g_3).

Any function with periods ω_1 and ω_2, such as $\wp$, has values that repeat in each parallelogram in the tessellation of the plane $\mathbb{C}$ shown in Figure 9. We assume that ω_1 and ω_2 lie in different directions from O, so that their integer combinations $m\omega_1 + n\omega_2$ form a *lattice* of parallelograms. The points marked by stars, which are "equivalent modulo the lattice," form a set on which all values of the function are the same.

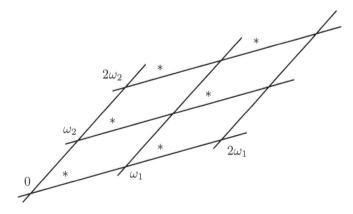

FIGURE 9. Lattice-equivalent points

Thus $\wp$ can be viewed as a function on the surface whose "points" are the classes of lattice-equivalent points

$$z + \omega_1\mathbb{Z} + \omega_2\mathbb{Z} = \{z + m\omega_1 + n\omega_2 : m, n \in \mathbb{Z}\}.$$

We call the surface the *quotient* $\mathbb{C}/\Gamma$ of the plane $\mathbb{C}$ by the group Γ of translations $z \mapsto z + m\omega_1 + n\omega_2$ for $m, n \in \mathbb{Z}$. This Riemann surface, not surprisingly, is a torus, obtained by identifying (or "pasting") opposite sides of the parallelogram with vertices $0, \omega_1, \omega_2, \omega_1 + \omega_2$, as shown in Figure 10. Thus, we can also arrive at meromorphic functions on the torus by starting with suitably *periodic* functions in the plane $\mathbb{C}$. In the years 1880–1882 (when the Dedekind-Weber paper was awaiting

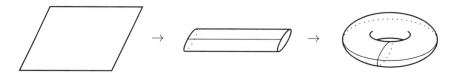

FIGURE 10. Construction of a torus by pasting

publication) a remarkable approach to meromorphic functions on surfaces of genus $p > 1$ was discovered through an exploration of *non-Euclidean periodicity*, on the half-plane $\{z : \text{Im}(z) > 0\}$ or the disk $\{z : |z| < 1\}$.

Isolated examples of functions with striking periodic behavior were discovered before 1880, but their periodicity did not have a context or a name. The first picture of this new kind of periodicity to appear in print was given by Schwarz (1872), and it exhibits the periodicity of a function now known as a Schwarz triangle function. The periodicity is indicated in Figure 11 via a tessellation of the disk by curvilinear triangles, in each of which the function repeats its values.

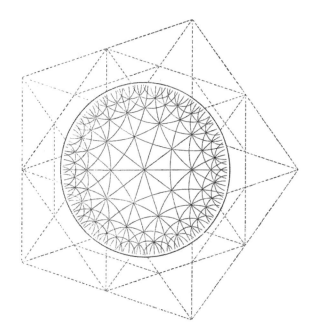

FIGURE 11. The Schwarz tessellation

The triangles are of course not congruent in the euclidean sense, but in 1880 Poincaré realized that they are *congruent in the sense of the non-Euclidean geometry* of Bolyai and Lobachevsky. Following this discovery, Poincaré (1882) brought non-Euclidean geometry into mainstream mathematics, and unveiled a new view of functions on Riemann surfaces—as functions with non-Euclidean periodicity.

The relationship of the disk to a Riemann surface $\mathcal{S}$ of genus $p > 1$ is like that of the plane to the torus: $\mathcal{S}$ is a quotient of the disk by a discrete group of (non-Euclidean) translations, and the disk is tessellated by congruent copies of a polygon obtained by suitably cutting the surface. The simplest case is the surface of genus 2, which can be cut along the curves shown in Figure 12 to form an octagon whose corner angles sum to 2π.

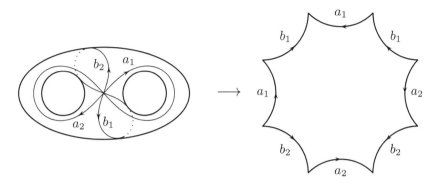

FIGURE 12. Dissection of genus 2 surface

The natural home of such an octagon is the non-Euclidean plane, which can be modeled by the unit disk with circular arcs orthogonal to the boundary circle as "lines." In particular, we can arrange eight such arcs so as to form an octagon with corner angles $\pi/4$, and (non-Euclidean) translations of this octagon fill the whole disk as shown in Figure 13.

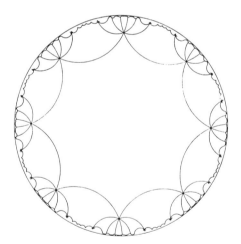

FIGURE 13. Tessellation of the disk by octagons

We can then recover the surface $\mathcal{S}$ of genus 2 as the quotient of the disk by the group Γ of all translations that map the tessellation onto itself. Finally, meromorphic functions on $\mathcal{S}$ correspond to meromorphic functions f on the disk with periodicity given by Γ, that is, functions with the property that

$$f(z) = f(g(z)), \quad \text{for any translation } g \in \Gamma.$$

Thus we are reduced to the problem of constructing meromorphic functions f on the disk with periodicity given by Γ. This problem was solved by Poincaré (1883), using insights from the theory of elliptic functions. The solution builds all elements $g \in \Gamma$ into f, though not so simply as Eisenstein (1847) built Euclidean translations into the definition of a doubly-periodic function. We omit the details.

Poincaré's construction of functions with non-Euclidean periodicity (*Fuchsian functions*, or *automorphic functions* as they became known) was a spectacular extension of the theory of elliptic functions to surfaces of higher genus. At the same time, it showed that elliptic functions and the torus are special, being associated with Euclidean geometry, and that the general Riemann surface should be viewed as non-Euclidean. However, like Riemann, Poincaré assumed an unproved theorem— the so-called *uniformization theorem*—to give his results their most general form.

Just as Riemann needed an assumption (the Dirichlet principle) to prove that any surface of genus 0 is isomorphic to the sphere $\mathbb{C} \cup \{\infty\}$, Poincaré needed to assume that any Riemann surface of genus $p > 1$ is isomorphic to a quotient of the disk by a discrete group of non-Euclidean translations. The uniformization theorem encompasses both these theorems by stating that any *simply connected* Riemann surface is isomorphic to either the sphere $\mathbb{C} \cup \{\infty\}$, the Euclidean plane $\mathbb{C}$, or the unit disk $\{z : |z| < 1\}$. Uniformization also extends the known parameterization theorems for algebraic curves of genus 0 (parameterization by rational functions) and genus 1 (parameterization by elliptic functions), because it implies that any algebraic curve of genus > 1 may be parameterized by automorphic functions.

Poincaré's work was clearly a brilliant extension of Riemann's ideas, but it remained under a cloud for some time because of doubts about its rigor. Indeed his papers of 1882 and 1883 were not published by the leading journals of the time; instead, they played a big part in establishing the new journal *Acta Mathematica*.[4] The doubts were eventually dispelled with proofs of the uniformization theorem by Poincaré (1907) and Koebe (1907), and by the rigorous theory of Riemann surfaces developed by Weyl (1913). Thus, it took more than 50 years for the analytic theory of algebraic curves to be fully accepted, and perhaps this worked to the advantage of algebra and algebraic geometry. The algebraic restructuring of Riemann's concepts by Dedekind and Weber may not have taken place had an analytic alternative been available, and we might thereby have missed a radically new insight into the nature of algebraic curves. The first textbook development of the algebraic theory was Hensel and Landsberg (1902), and then only with a reversion to certain ideas from analysis, such as infinite series expansions. For an English version of the Hensel-Landsberg approach, with some simplifications, see Bliss (1933).

Even today, it is a little strange to think that number theory inspired the first complete and rigorous proof of the Riemann-Roch theorem. Nevertheless, if one looks at the proof of the theorem in any modern book one will see that it involves things called "divisors." A divisor is nothing but a finite set of points with attached integer multiplicities—like a set of zeros and poles—yet it sounds as if the concept comes from number theory. In the next sections we explain how this happened.

---

[4]Looking back to his founding of *Acta Mathematica*, Mittag-Leffler (1923) recalled:

> Kronecker, for example, expressed to me via a mutual friend his regret that the journal seemed bound to fail, without help, through publishing a work so incomplete, unripe, and obscure.

7. Origins of Algebraic Number Theory

> It is greatly to be lamented that this virtue of the real numbers [i.e., the rational integers] to be decomposable into prime factors, always the same ones for a given number, does not also belong to the complex numbers; were this the case, the whole theory, which is still laboring under such difficulties, could easily be brought to its conclusion. For this reason, the complex numbers we have been considering seem imperfect, and one may well ask whether one ought to look for another kind which would preserve the analogy with the real numbers ...

> Translation by Weil (1975) from Kummer (1844).

The concept of divisibility has been fundamental in number theory for more than 2000 years. Euclid's *Elements*, Book VII, introduces the Euclidean algorithm in Proposition 1, and shows that it yields the greatest common divisor of two numbers in Proposition 2. Eventually Euclid deduces, in Proposition 30, that if a prime p divides a product ab then p divides a or p divides b. This prime divisor property easily yields what we now call the *fundamental theorem of arithmetic*: each natural number has unique prime factorization (up to the order of factors).

Unique prime factorization was first explicitly stated as a theorem in Gauss (1801), but before then it was frequently assumed without comment. Indeed, unique prime factorization was sometimes assumed for numbers seemingly far from the natural numbers. One of the first instances was the spectacular proof of Euler (1770), p. 401, that the only natural number solution of

$$y^3 = x^2 + 2$$

is $x = 5$, $y = 3$, a result that had been claimed by Fermat. To prove this result, Euler factorized the right side of the equation, obtaining

$$y^3 = (x + \sqrt{-2})(x - \sqrt{-2}),$$

and then worked with numbers of the form $a + b\sqrt{-2}$ (for integers a and b) as if they were ordinary integers. In particular, he argued that $x + \sqrt{-2}$ and $x - \sqrt{-2}$ are relatively prime and therefore (apparently assuming unique prime factorization) that each is a cube, because their product is a cube. But if

$$x + \sqrt{-2} = (a + b\sqrt{-2})^3$$

it follows that

$$x = a^3 - 6ab^2 = a(a^2 - 6b^2), \quad 1 = 3a^2b - 2b^3 = b(3a^2 - 2).$$

The only integer solutions of the latter equation are $a = \pm 1$, $b = 1$, of which only $a = -1$, $b = 1$ give a natural number solution, namely $x = 5$.

This is magnificent, but is it number theory?

Evidently, we need to rebuild the theory of divisibility and primes for new types of numbers, such as those of the form $a + b\sqrt{-2}$. The first to take steps in this direction was Gauss (1832), who worked out the basic theory of what we now call the *Gaussian integers*, $\mathbb{Z}[i]$. These are the numbers of the form $a + bi$, where a and b are integers and $i = \sqrt{-1}$.

We now sketch the divisibility theory of $\mathbb{Z}[i]$, since it admits a visual interpretation that will be helpful in other cases. The only result we really need to prove is the *division property* (sometimes called the "division algorithm"): *if α and $\beta \neq 0$*

are Gaussian integers, then there are Gaussian integers μ and ρ such that

$$\alpha = \mu\beta + \rho, \quad where \quad |\rho| < |\beta|.$$

If we can prove this, then there is a Euclidean algorithm, and unique prime factorization in $\mathbb{Z}[i]$ follows as it does for ordinary integers.

Well, for any Gaussian integer $\beta \neq 0$, consider the multiples $\mu\beta$ of β by all the Gaussian integers μ. These are the integer combinations of the perpendicular vectors β and $i\beta$, which form squares of side length $|\beta|$, as shown in Figure 14.

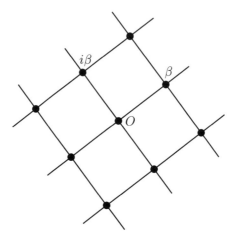

FIGURE 14. Multiples of β in $\mathbb{Z}[i]$

The Gaussian integer α falls in one of these squares, so if $\mu\beta$ is the nearest corner we have

$$|\rho| < |\beta|, \quad where \quad \rho = \alpha - \mu\beta,$$

because the distance of a point in a square from the nearest corner is less than the side length. This establishes the division property, and hence unique prime factorization in $\mathbb{Z}[i]$.

The argument is similar for the set $\mathbb{Z}[\sqrt{-2}]$ of numbers of the form $a + b\sqrt{-2}$ considered by Euler. The multiples $\mu\beta$ of such a number β form a grid of rectangles, with sides of length $|\beta|$ and $|\beta|\sqrt{2}$, but it remains true that the distance of a point α from the nearest corner is less than $|\beta|$. Thus the division property holds for $\mathbb{Z}[\sqrt{-2}]$, hence a Euclidean algorithm and unique prime factorization, so the main assumption of Euler's proof is valid. (The other assumption, about relative primality, is also easy to justify.)

Many results about ordinary integers are most easily proved by passing to "quadratic integers," such as $\mathbb{Z}[i]$, and appealing to unique prime factorization. For example, consider the prime 797, which is the sum of two squares $26^2 + 11^2$. Is this sum of squares unique? Yes, because if $797 = a^2 + b^2$ we have $797 = (a+bi)(a-bi)$, which is a *prime* factorization because $|a + bi|^2 = |a - bi|^2$ is the ordinary prime $a^2 + b^2 = 797$ (so neither $a+bi$ nor $a-bi$ is a product of smaller Gaussian integers). Then, since Gaussian prime factorization is unique, so is the decomposition of 797 into a sum of squares.

Unfortunately, there is trouble not far ahead. If we wish to prove theorems about natural numbers of the form $a^2 + 5b^2$ by using the factorization

$$a^2 + 5b^2 = (a + b\sqrt{-5})(a - b\sqrt{-5})$$

we cannot assume unique prime factorization, because it *fails* for the set $\mathbb{Z}[\sqrt{-5}]$ of numbers of the form $a + b\sqrt{-5}$. The classical example, from Dedekind (1877), is

$$6 = 2 \cdot 3 = (1 + \sqrt{-5})(1 - \sqrt{-5}).$$

Each of the numbers 2, 3, $1 + \sqrt{-5}$, $1 - \sqrt{-5}$ is "prime," in the sense that none of them is a product of smaller numbers in $\mathbb{Z}[\sqrt{-5}]$, so we have nonunique prime factorization in $\mathbb{Z}[\sqrt{-5}]$. Gauss seemed to be aware that unique prime factorization could fail in cases like this; his theory of quadratic forms in Gauss (1801) seems designed to avoid such problems, sometimes at considerable expense.

However the first to note the phenomenon in print, and to declare that something should be done about it, was Kummer (1844) (see the quote at the beginning of this section). In a bold attempt to rescue unique prime factorization, Kummer introduced what he called "ideal numbers." The name was apparently motivated by the "ideal" elements that were then becoming accepted in geometry, such as points at infinity, but the *concept* drew inspiration from Jacobi's work on number theory in the 1830s, as has been shown by Lemmermeyer (2009). In Kummer's work, "ideal numbers" arise in a rather complicated way, and we will instead follow Dedekind (1877) and explain how they work in $\mathbb{Z}[\sqrt{-5}]$.

To reconcile the the factorizations

$$6 = 2 \cdot 3 = (1 + \sqrt{-5})(1 - \sqrt{-5})$$

with unique prime factorization, we have to believe that 2, 3, $1 + \sqrt{-5}$, $1 - \sqrt{-5}$ are *not* primes, after all. They must somehow split into smaller, "ideal," factors— but what are these ideal factors? A candidate that comes to mind is the greatest common divisor of 2 and $1 + \sqrt{-5}$. We cannot put our finger on any such number but, by borrowing an idea from classical number theory, we can describe the *set of multiples* of a greatest common divisor. In classical number theory, the multiples of $\gcd(a, b)$ are all the numbers of the form $ma + nb$. In $\mathbb{Z}[\sqrt{-5}]$, there is no actual number that can serve as $\gcd(2, 1 + \sqrt{-5})$, but the set

$$\{2\mu + (1 + \sqrt{-5})\nu : \mu, \nu \in \mathbb{Z}[\sqrt{-5}]\} = \{2m + (1 + \sqrt{-5})n : m, n \in \mathbb{Z}\}$$

is perfectly concrete, and it can serve as the set of "multiples of the ideal number $\gcd(2, 1 + \sqrt{-5})$."

Figure 15 shows what this set looks like. Its members are the black dots on the grid of points in $\mathbb{Z}[\sqrt{-5}]$. Notice that these black dots do *not* form a grid of rectangles, as they would if they were multiples of an actual number in $\mathbb{Z}[\sqrt{-5}]$—because the multiples of a number α are simply the rectangular grid $\mathbb{Z}[\sqrt{-5}]$, magnified by $|\alpha|$ and rotated through the argument of α.

The number $\gcd(2, 1 + \sqrt{-5})$ is "ideal," but the set of its multiples is perfectly real, and visible. When we admit "ideal numbers," via their sets of multiples, it turns out that there are "ideal primes," and that each number in $\mathbb{Z}[\sqrt{-5}]$ factorizes uniquely into ideal primes. This is how Kummer recovered unique prime factorization, and hence preserved the analogy with the ordinary integers.

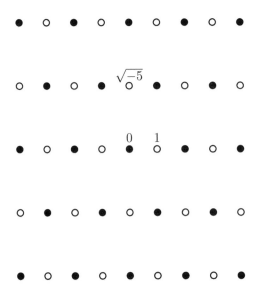

FIGURE 15. Multiples of the ideal number $\gcd(2, 1 + \sqrt{-5})$ in $\mathbb{Z}[\sqrt{-5}]$

8. Dedekind's Theory of Algebraic Integers

> The great success of Kummer's researches in the domain of circle division allows us to suppose that the same laws hold in *all* numerical domains .... I did not achieve the general theory ... until I abandoned the old formal approach and replaced it by another; a fundamentally simpler conception focussed directly on the goal. In the latter approach I need no concept more novel than that of Kummer's *ideal numbers*, and it is sufficient to consider a *system of actual numbers* that I call an *ideal*.
>
> Dedekind (1877), pp. 56–57.

Kummer was interested in a particular class of numbers, now known as *cyclotomic integers*, most famously for their application to Fermat's last theorem. He developed his theory of "ideal numbers" only for the cyclotomic integers, and it was left to Dedekind to develop a general theory of algebraic integers, and what we now know as the theory of ideals. Dedekind's first account of this theory appeared in Dedekind (1871), an appendix to Dirichlet's *Vorlesungen über Zahlentheorie*, which Dedekind edited. This did not attract as much interest as Dedekind hoped, and he produced a more down-to-earth exposition of the theory in Dedekind (1877). I have drawn on the latter version here.

From a handful of examples such as $\mathbb{Z}[i]$, $\mathbb{Z}[\sqrt{-2}]$, and Kummer's cyclotomic integers, Dedekind (1871) synthesized a general concept of *algebraic integer*. Then, by properly situating algebraic integers in the broader context of algebraic *numbers*, he was able to define ideal prime numbers and prove the uniqueness of ideal prime factorization. In outline, his train of thought was the following.

Begin with the *algebraic numbers*. A number α is algebraic if it is the root of an equation

$$a_n x^n + a_{n-1} x^{n-1} + \cdots + a_1 x + a_0 = 0, \quad \text{where} \quad a_0, a_1, \ldots, a_n \in \mathbb{Z}.$$

Among the algebraic numbers, those satisfying equations of this form in which $a_n = 1$ (*monic* equations) are called *algebraic integers*. It can be proved (not quite trivially) that the algebraic numbers are closed under the operations $+, -, \times, \div$ (by a nonzero number), and hence they form a field, and that the algebraic integers are closed under the operations $+, -, \times$, and hence they form a ring. The concept of algebraic integer is compatible with the ordinary concept of integer because any rational solution of a monic equation is an ordinary integer. This fact was already mentioned by Gauss (1801), article 11.

For these reasons, and others that will appear below, the concept of algebraic integer is a good generalization of the ordinary integer concept. This is not completely obvious, because there are algebraic integers that do not "look integral." One might think that the cube root of 1,

$$\zeta_3 = \frac{-1 + \sqrt{-3}}{2},$$

should not be regarded as integer because of its "fractional" appearance. Yet it is a root of the monic equation, $x^2 + x + 1 = 0$. And in fact it is better to work in the ring

$$\mathbb{Z}[\zeta_3] = \{a + b\zeta_3 : a, b \in \mathbb{Z}\}$$

than in the ring

$$\mathbb{Z}[\sqrt{-3}] = \{a + b\sqrt{-3} : a, b \in \mathbb{Z}\},$$

because $\mathbb{Z}[\zeta_3]$ has unique prime factorization and $\mathbb{Z}[\sqrt{-3}]$ does not.

However, it is *not* a good idea to work with the ring of all algebraic integers, because it cannot possibly have unique prime factorization. This is because any algebraic integer α has a factorization

$$\alpha = \sqrt{\alpha}\sqrt{\alpha},$$

and $\sqrt{\alpha}$ is also an algebraic integer. Dedekind (1871) saw that the way to avoid this problem is to work in an *algebraic number field F of finite degree over* $\mathbb{Q}$, and with the ring of algebraic integers in F. Any such field has the form[5]

$$\mathbb{Q}(\alpha) = \{p(\alpha)/q(\alpha) : p, q \text{ polynomials with integer coefficients}\},$$

for some algebraic number α. The degree of α (that is, the degree of the minimal polynomial satisfied by α) is called the *degree of* $F = \mathbb{Q}(\alpha)$, $\deg(F)$.

Each algebraic integer β in F has a minimal polynomial

$$x^n + b_{n-1} x^{n-1} + \cdots + b_1 x + b_0,$$

where $n \leq \deg(F)$, and $b_0, \ldots, b_{n-1}$ are ordinary integers. The ordinary integer $(-1)^n b_0$ is called the *norm* of β, $N(\beta)$. $N(\beta)$ is the product of β with all the other roots of the minimal polynomial, called the *conjugates* of β. For example,

---

[5]Dedekind actually defines a field F of finite degree n over $\mathbb{Q}$ more generally, as a field of dimension n as a vector space over $\mathbb{Q}$. But it is true that there is always a *primitive element* α such that $F = \mathbb{Q}(\alpha)$. This result, essentially due to Galois, also applies to the function fields that appear later, so we will also suppose each of them to be generated by a primitive element.

a Gaussian integer $\beta = a + bi$ has one conjugate, the ordinary complex conjugate $\overline{\beta} = a - bi$, and

$$N(\beta) = (a + bi)(a - bi) = a^2 + b^2 = |\beta|^2.$$

The familiar property of absolute value, $|\alpha\beta| = |\alpha||\beta|$, generalizes to the *multiplicative property of norm*:

$$N(\alpha\beta) = N(\alpha)N(\beta).$$

It follows from the multiplicative property that, if $\gamma = \alpha\beta$, then we have $N(\gamma) = N(\alpha)N(\beta)$. This reduces certain questions about divisibility of algebraic integers to questions about divisibility of ordinary integers. For example, if α divides γ, then $N(\alpha)$ divides $N(\gamma)$. It follows that the process of factorizing an algebraic integer γ must eventually terminate with a factorization

$$\gamma = \alpha_1\alpha_2\cdots\alpha_k,$$

where $N(\alpha_1), N(\alpha_2), \ldots, N(\alpha_k)$ are ordinary primes.

Thus any integer γ of F has a factorization into integers $\alpha_1, \alpha_2, \ldots, \alpha_n$ that are *irreducible* in the sense that α_i is not the product of integers of smaller norm.

We know from the example

$$6 = 2 \cdot 3 = (1 + \sqrt{-5})(1 - \sqrt{-5})$$

in $\mathbb{Z}[\sqrt{-5}]$, which is the ring of integers of $\mathbb{Q}(\sqrt{-5})$, that factorization into irreducibles is not always unique. But we also know that unique prime factorization can be recovered, in this case, by further splitting into "ideal numbers." Dedekind generalized this recovery program with his concept of *ideals*. An *ideal* $\mathfrak{a}$ is a set of integers of a field F with the following closure properties:

1. If $\alpha \in \mathfrak{a}$ and $\alpha' \in \mathfrak{a}$ then $\alpha + \alpha' \in \mathfrak{a}$ and $\alpha - \alpha' \in \mathfrak{a}$.
2. If $\alpha \in \mathfrak{a}$ and μ is any integer of F, then $\mu\alpha \in \mathfrak{a}$.

These properties capture the idea of a "set of integer multiples" of an integer, actual or ideal.

If α is an integer of F then its set of multiples

$$(\alpha) = \{\mu\alpha : \mu \text{ an integer of } F\}$$

certainly has the properties of an ideal. We call (α) the *principal ideal* generated by α. The set in $\mathbb{Z}[\sqrt{-5}]$,

$$\{2\mu + (1 + \sqrt{-5})\nu : \mu, \nu \in \mathbb{Z}[\sqrt{-5}]\} = \{2m + (1 + \sqrt{-5})n : m, n \in \mathbb{Z}\},$$

also has the required closure properties, so it is also an ideal. But, as we have seen, this ideal is *not* the set of integer multiples of an actual integer in $\mathbb{Q}(\sqrt{-5})$—it is a *nonprincipal* ideal.

If $\mathfrak{a}$ and $\mathfrak{b}$ are ideals, Dedekind (1871) defines their *product* by

$$\mathfrak{a}\mathfrak{b} = \{\alpha_1\beta_1 + \cdots + \alpha_m\beta_m : \alpha_1, \ldots, \alpha_m \in \mathfrak{a} \text{ and } \beta_1, \ldots, \beta_m \in \mathfrak{b}\}.$$

This definition agrees with the natural product $(\alpha)(\beta) = (\alpha\beta)$ when $\mathfrak{a} = (\alpha)$ and $\mathfrak{b} = (\beta)$ are principal ideals. Finally, Dedekind (1871) says that $\mathfrak{b}$ *divides* $\mathfrak{a}$ if $\mathfrak{b} \supseteq \mathfrak{a}$ ("to divide is to contain"). This also agrees with the natural concept of division for principal ideals. For example, 2 divides 6, so naturally (2) divides (6), and indeed $(2) \supseteq (6)$ because

$$(2) = \{0, \pm 2, \pm 4, \pm 6, \ldots\} \supseteq \{0, \pm 6, \pm 12, \ldots\} = (6).$$

Up to this point, the theory is a straightforward extension of the usual divisibility concept to ideals, if one views ideals as sets of multiples. But, as Dedekind (1877), §23, points out:

> It is very easy to see (§22,1) that any product of $\mathfrak{a}$ by an ideal $\mathfrak{b}$ is divisible by $\mathfrak{a}$, but it is by no means easy to show the converse, that each ideal divisible by $\mathfrak{a}$ is the product of $\mathfrak{a}$ by an ideal $\mathfrak{b}$.

Eventually, Dedekind overcomes the difficulty by generalizing the theory of divisibility to more general subsystems of integers of F called *orders*, and unique prime ideal factorization follows.

We will not go into further detail, because the details are somewhat different in the case of algebraic function fields, as Dedekind and Weber point out. We only wish to emphasize how important it was to be aware of the notion of ideal. The theory of algebraic functions would not have been discovered without this awareness of ideals, or something similar.

There is in fact an alternative to the concept of ideal, namely, the concept of *divisor* developed by Kronecker. For the full story, see Edwards (1980). Kronecker's work overlaps with Dedekind's and in fact his major paper, Kronecker (1882), appears in the same issue of the journal containing the Dedekind-Weber paper (probably not by accident,[6] since Kronecker was an editor of the journal). Kronecker's paper surpasses the Dedekind-Weber paper in some ways, by considering functions of several variables, but it falls short on specifics and applications, such as proofs of the Abel and Riemann-Roch theorems. For this reason, and its greater readability, the Dedekind-Weber paper has been far more influential. Nonetheless, we must give credit to Kronecker for his divisor concept. The word "divisor" is the one that has passed into algebraic geometry today, in place of what Dedekind and Weber called a "polygon." In the next section we explain how the concept of divisibility made its way from algebraic number theory to algebraic geometry.

9. Number Fields and Function Fields

> Dedekind and Weber propose to give algebraic proofs of all of Riemann's algebraic theorems. But their remarkable originality (which in all the history of algebraic geometry is only scarcely surpassed by that of Riemann) leads them to introduce a series of ideas that will become fundamental in the modern era.
>
> Dieudonné (1985), p. 29.

If we had to explain the nature of algebraic number theory in a nutshell, we could say that it is the result of generalizing the theory of $\mathbb{Z} \subseteq \mathbb{Q}$ (the "rational integers") to the integers in an extension field $\mathbb{Q}(\alpha)$ of $\mathbb{Q}$, where α is an algebraic number. In particular, one tries to generalize the theory of divisibility and primes, with a view to retaining, or recovering, unique prime factorization. The search for suitable "primes" leads to the discovery of the key concept of ideal.

The 19th-century theory of algebraic functions begins with the analogous theory of polynomials in the field $\mathbb{C}(x)$ of rational functions of x with complex coefficients, and it similarly studies extensions of $\mathbb{C}(x)$. So first we should see to what extent the polynomials behave like "integers" in the field of rational functions.

---

[6]Indeed, Kronecker delayed publication of the Dedekind-Weber paper for over a year, until his paper could appear at the same time. See Edwards (1980), p. 370.

Certainly, the polynomials form a ring. Also, it has been known since Stevin (1585) that they have a division property and hence they admit a Euclidean algorithm. The division property takes the following form: if $f(x)$ and $g(x) \neq 0$ are polynomials, then there are polynomials $q(x)$ and $r(x)$ ("quotient" and "remainder") such that

$$f(x) = g(x)q(x) + r(x), \quad \text{where} \quad \deg(r) < \deg(g).$$

Thus the remainder $r(x)$ is "smaller" than $g(x)$ when measured by its degree. The Euclidean algorithm therefore terminates, and yields the gcd of polynomials $f(x)$ and $g(x)$, in a number of steps bounded by the degree of g. It follows by the usual steps that unique prime factorization holds for the ring $\mathbb{C}[x]$ of polynomials in $\mathbb{C}(x)$. To be precise, prime factorization is unique up to nonzero constant factors. Moreover, we can say exactly what the prime polynomials are: by the fundamental theorem of algebra, they are the linear polynomials $x - c$, for complex numbers c.

Now the field $\mathbb{C}(x)$ of rational functions is, as we saw in Section 4, the field of meromorphic functions on the sphere $\mathbb{C} \cup \{\infty\}$. The primes $x - c$ of this field correspond to the points c of $\mathbb{C}$. So, in a sense, we can recover the points of the surface on which the field $\mathbb{C}(x)$ "lives" from the primes of the field itself. (Admittedly, we are still missing the point ∞, but presumably we can squeeze it out of $\mathbb{C}(x)$ somehow.)

This seemingly retrograde idea—defining a Riemann surface from the functions on it—is one of the most original and prescient ideas of the Dedekind-Weber paper. But before we can explain how it works, we must say more about function fields, in order to show how a function field may be plausibly related to a Riemann surface in the first place. As mentioned in Section 4, the connection arises from Riemann's discovery that the meromorphic functions on a Riemann surface are algebraic, that is, solutions of polynomial equations. It is also clear that the set of all meromorphic functions on a given Riemann surface forms a field, because it is obviously closed under the operations of $+$, $-$, $\times$, and $\div$ (by a nonzero function).

Now, if we pursue that idea of constructing algebraic function fields in analogy with algebraic number fields, then we should extend $\mathbb{C}(x)$ to $\mathbb{C}(x)(y)$, where y (a "primitive element" for the field) satisfies an equation of the form

$$a_n y^n + a_{n-1} y^{n-1} + \cdots + a_1 y + a_0 = 0, \quad \text{where} \quad a_0, a_1, \ldots, a_n \in \mathbb{C}(x).$$

The elements of $\mathbb{C}(x)$ are rational functions, that is, quotients of polynomials. By multiplying by their common denominator we can therefore rewrite the equation satisfied by y in the form

$$f(x, y) = 0,$$

where $f(x, y)$ is a polynomial. In other words, y is defined by the equation of an algebraic curve, and hence of a Riemann surface. The function field generated by y consists of the rational functions in y,

$$\mathbb{C}(x)(y) = \{p(y)/q(y) : p, q \text{ polynomials with coefficients in } \mathbb{C}(x)\}$$
$$= \{\text{rational functions in } x, y \text{ such that } f(x, y) = 0\}$$
$$= \mathbb{C}(x, y)/(f(x, y)),$$

which is called the field of rational functions *on* the curve $f(x, y) = 0$, or on the Riemann surface $f(x, y) = 0$. Thus a finite-degree extension of $\mathbb{C}(x)$ is naturally viewed as the field of rational functions on a Riemann surface.

Exactly why the field Σ rational functions in x, y such that $f(x, y) = 0$ should be called "the rational functions on the curve $f(x, y) = 0$" is seldom explained. One of the few authors who bothers to do so is Walker (1950), p. 132 (by "a point of f" he means a point of the curve $f(x, y) = 0$):

> If $g_1(x, y)/h_1(x, y)$ and $g_2(x, y)/h_2(x, y)$ are rational functions ... and (a, b) is any point of f at which these functions have values, then these values will be equal if $g_1(\xi, \eta)/h_1(\xi, \eta) = g_2(\xi, \eta)/h_2(\xi, \eta)$. Conversely, if this last equality does not hold, then $f(x, y)$ does not divide $g_1(x, y)h_2(x, y) - g_2(x, y)h_1(x, y)$, and there are points of f at which the two rational functions assume different values. In other words, as far as the points of f are concerned the rational functions behave like elements of Σ.

We now pause to look at some examples of finite-degree extensions of $\mathbb{C}(x)$.

A genus 0 example. The rational functions on $x^2 + y^2 = 1$.

This curve is just the complex version of the circle, which is well known to have a parameterization by rational functions:

$$x = \frac{2t}{1 + t^2}, \quad y = \frac{1 - t^2}{1 + t^2}.$$

(This result is equivalent to Euclid's parameterization of Pythagorean triples, given in a lemma following Prop. 28 in Book X of the *Elements*.)

A rational function of x and y is therefore a rational function of t, so our function field in this case is a subfield of $\mathbb{C}(t)$. In fact our field is *exactly* $\mathbb{C}(t)$, because

$$\frac{x}{1 + y} = t,$$

so any rational function of t is a rational function of x and y.

Thus, the field of rational functions on $x^2 + y^2 = 1$ is the same as the field of rational functions on $\mathbb{C} \cup \{\infty\}$. If we hope to recover each Riemann surface from the field of rational functions on it, the Riemann surface $x^2 + y^2 = 1$ therefore needs to be the "same" as the sphere $\mathbb{C} \cup \{\infty\}$ in some sense. The equations above show that $x^2 + y^2 = 1$ and $\mathbb{C} \cup \{\infty\}$ are indeed the "same" in the following strong sense: they are *birationally equivalent*; that is, there is a one-to-one correspondence $(x, y) \leftrightarrow t$ that is rational in both directions. $\qquad\square$

This notion of equivalence was discovered by Riemann (1851), who showed a much stronger result, already alluded to in Section 4. Assuming the Dirichlet principle, he showed that any Riemann surface of genus zero is birationally equivalent to the sphere $\mathbb{C} \cup \{\infty\}$. A simpler fact about birational equivalence is that *algebraic curves $f(x, y) = 0$ and $g(x, y) = 0$ are birationally equivalent if and only if their function fields are isomorphic*, which is an easy generalization of the example above.

A genus 1 example. The rational functions on $y^2 = 1 - x^4$.

The Riemann surface $y^2 = 1 - x^4$ is *not* birationally equivalent to the sphere, because it has no parameterization $x = f(t)$, $y = g(t)$ by rational functions f and g. This is the result foreseen by Jakob Bernoulli (1704) that we mentioned in Section 2. It is worth saying a little more about it now, because it nicely underlines the analogy between polynomials and integers.

Suppose on the contrary that there are rational functions $x(t)$ and $y(t)$ with

$$y(t)^2 = 1 - x(t)^4.$$

Writing $x(t) = p(t)/r(t)$ and $y(t) = q(t)/r(t)$ as quotients of polynomials with a common denominator we get the equation

$$q(t)^2 r(t)^2 = r(t)^4 - p(t)^4,$$

and hence

$$s(t)^2 = r(t)^4 - p(t)^4,$$

for certain polynomials $p(t)$, $r(t)$, and $s(t)$.

Now, as Jakob Bernoulli noticed, there is a theorem of Fermat that the equation $s^2 = r^4 - p^4$ is impossible for *nonzero integers* p, r, s. This does not rule out the equation for nontrivial *polynomials* p, r, s, but Bernoulli was nevertheless on the right track. One has only to imitate Fermat's argument, using polynomials in place of integers. The two domains are sufficiently similar that the arguments carry over.

For example, the starting point of Fermat's argument is essentially the fact mentioned in the previous example: if X, Y are rational numbers such that

$$X^2 + Y^2 = 1,$$

then we can write

$$X = \frac{2T}{1+T^2}, \quad Y = \frac{1-T^2}{1+T^2}, \quad \text{for the rational number } T = \frac{X}{1+Y}.$$

In our case we have rational functions $X(t), Y(t)$ such that $X(t)^2 + Y(t)^2 = 1$, and exactly the same calculation shows that

$$X(t) = \frac{2T(t)}{1+T(t)^2}, \quad Y(t) = \frac{1-T(t)^2}{1+T(t)^2}, \quad \text{for the rational } \textit{function } T(t) = \frac{X(t)}{1+Y(t)}.$$

The rest of the argument may be similarly rewritten, using divisibility of polynomials in place of divisibility of numbers where appropriate.

Thus the polynomial version of Fermat's argument shows that the Riemann surface $y^2 = 1 - x^4$ is not birationally equivalent to the sphere. It follows, by the remark after the previous example, that the field of rational functions on $y^2 = 1 - x^4$ is not isomorphic to $\mathbb{C}(x)$.[7]

The function field of $y^2 = 1 - x^4$ therefore reflects the difference between this genus 1 curve and the genus 0 curve $\mathbb{C} \cup \{\infty\}$. In fact, it reflects more than the difference in genus, because the genus 1 curves actually fall into infinitely many

---

[7]The function field of $y^2 = 1 - x^4$ in fact equals $\mathbb{C}(\mathrm{sl}(t), \mathrm{sl}'(t))$, where sl is the lemniscatic sine function defined, as in Section 2, by

$$u = \mathrm{sl}^{-1}(x) = \int_0^x \frac{dt}{\sqrt{1-x^4}}.$$

This is because the curve $y^2 = 1 - x^4$ has the parameterization

$$x = \mathrm{sl}(u), \quad y = \mathrm{sl}'(u).$$

The definition of sl^{-1} of course gives $x = \mathrm{sl}(u)$, and differentiation gives

$$\frac{du}{dx} = \frac{1}{\sqrt{1-x^4}} = \frac{1}{y},$$

whence

$$\mathrm{sl}'(u) = \frac{dx}{du} = y.$$

Thus the rational functions on $y^2 = 1 - x^4$ are generated from the elliptic functions sl and sl$'$.

birational equivalence classes. This follows from a theorem of Salmon (1851), if one suitably reinterprets Salmon's result. He stated his result in terms of projective equivalence, which happens to be the same as birational equivalence, and also as a result about curves of degree 3 rather than curves of genus 1. The curve $y^2 = 1 - x^4$ is not itself of degree 3, of course, but it is birationally equivalent to the cubic curve

$$Y^2 = 4X^3 - 6X^2 + 4X - 1$$

under the substitution

$$X = \frac{1}{1-x}, \quad Y = \frac{y}{(1-x)^2}. \qquad \Box$$

The correspondence between birational equivalence classes of curves and iso-morphism classes of their function fields is now taken for granted—so much so that certain theorems that were discovered as results about birational equivalence are now stated (without comment) as theorems about function fields. A well-known example is the so-called *Lüroth's theorem*. The original statement of the theorem, in Lüroth (1875), is that any curve parameterized by rational functions can be pa-rameterized *bijectively* by rational functions (with the exception of finitely many points, for example, if the curve has self-intersections). A typical modern statement of the theorem is that any subfield of $\mathbb{C}(z)$ containing more than $\mathbb{C}$ is isomorphic to $\mathbb{C}(z)$.

The equivalence of these two statements is not obvious, though not very hard to prove either. For an elementary proof that the modern form of the theorem implies the original form, see Shafarevich (1994), pp. 9–10.

10. Algebraic Functions and Riemann Surfaces

> These prime ideals correspond to the linear factors in the theory of polynomials. On this basis one attains a completely precise and gen-eral definition of a "point of a Riemann surface," i.e., a complete system of numerical values that can be consistently attached to the functions of the field.
>
> Dedekind and Weber (1882), Introduction.

The basic idea of Dedekind and Weber is very natural: a point a on a Riemann surface $\mathcal{S}$ gives a value $f(a)$ to each rational function f on $\mathcal{S}$, and the values given to different functions f and g are *consistent* in the sense that

- The value given to a constant function c is c.
- The value given to $f + g$ is $f(a) + g(a)$.
- The value given to $f - g$ is $f(a) - g(a)$.
- The value given to $f \times g$ is $f(a) \times g(a)$.
- The value given to $f \div g$ is $f(a) \div g(a)$.

When the values include ∞, as they necessarily do for rational functions, one has to make conventions such as $1/\infty = 0$ and $1/0 = \infty$, but this is not a serious problem. The basic question is: does each consistent assignment of values to the rational functions on a surface $\mathcal{S}$ arise from a unique point of $\mathcal{S}$? If so, we can reverse the original idea: namely, start with a function field F, and *define* points of a surface $\mathcal{S}_F$ for which F is the field of rational functions. Just say: a point $P \in \mathcal{S}_F$ *is* an assignment of values (from $\mathbb{C} \cup \{\infty\}$) to the functions in F that satisfies the consistency conditions above.

A genus 0 example. The rational function field $\mathbb{C}(x)$.

The idea that each point comes uniquely from an assignment of values to functions is certainly valid when the function field is $\mathbb{C}(x)$. In this case a consistent assignment of values to the functions in $\mathbb{C}(x)$ is determined by the values assigned to the prime functions $x - c$, where $c \in \mathbb{C}$. To be consistent with subtraction and the values of constant functions, each function $x - c$ must be given the value $a - c$, for some fixed $a \in \mathbb{C} \cup \{\infty\}$. But this is exactly the assignment given by the point $x = a$. Thus, for the function field of the surface $\mathcal{S} = \mathbb{C} \cup \{\infty\}$, each consistent assignment of values arises from a point $a \in \mathcal{S}$. $\qquad\qquad\square$

We can also see that any point P' of the surface $\mathcal{S}$ for an algebraic function field F lies "over" a point of the sphere $\mathbb{C} \cup \{\infty\}$. Why? Because F is a field containing $\mathbb{C}(x)$ and P' is a function $F \to \mathbb{C} \cup \{\infty\}$. Thus, if P is simply the restriction of the map P' to the subfield $\mathbb{C}(x)$, then P is a point of the surface for $\mathbb{C}(x)$, which corresponds to a point of $\mathbb{C} \cup \{\infty\}$, as just explained.

Moreover, if F is a field of degree n, then there are in general n points P' over a given point P (so the surface is an "n-sheeted covering" of the sphere). If P corresponds to the value $x = x_0$, consider a primitive element $y \in F$, which satisfies an equation[8]

$$y^n + a_{n-1}(x)y^{n-1} + \cdots + a_1(x)y + a_0(x) = 0,$$

where $a_0(x), \ldots, a_{n-1}(x) \in \mathbb{C}(x)$. The n points over P are the n assignments of values to the functions in F rising from the n values of y satisfying the equation

$$y^n + a_{n-1}(x_0)y^{n-1} + \cdots + a_1(x_0)y + a_0(x_0) = 0.$$

Thus the surface defined by Dedekind and Weber makes a good start in capturing the properties of a Riemann surface, and we can imagine identifying branch points, and so on. But where do ideals come in?

When we extend $\mathbb{C}(x)$ to an algebraic function field F, the analogy with algebraic number fields suggests that we should first define the "integers" of F, then the ideals. Nowadays, one generally skips the "integers" and goes straight to ideals, and beyond them to "divisors." We come to divisors shortly, but we will follow Dedekind and Weber by defining the "integers" first. The "integers" of $\mathbb{C}(x)$ are of course the polynomials in x, so the ring $\mathfrak{o}_F$ of "integers" of F should be the *integral closure* of the ring $\mathbb{C}[x]$ of polynomials with complex coefficients. This ring consists of the functions $y \in F$ satisfying equations of the form

$$y^n + b_{n-1}y^{n-1} + \cdots + b_1 y + b_0 = 0,$$

where $b_0, \ldots, b_{n-1} \in \mathbb{C}[x]$. Dedekind and Weber call these *ganz Functionen*, which we will translate as *integral algebraic functions*. They show in §3 that any element of F is a quotient of elements of $\mathfrak{o}_F$, in fact the quotient of an element of $\mathfrak{o}_f$ by a polynomial.

Ideals in $\mathfrak{o}_F$ are then defined as in any ring, by the closure properties given in Section 7. A *prime* ideal $\mathfrak{p}$ is divisible only by itself and $(1) = \mathfrak{o}_F$, so (since "to divide is to contain") $\mathfrak{p}$ is *maximal*. That is, if y is any integral algebraic function in $\mathfrak{o}_F \setminus \mathfrak{p}$, the only ideal containing both y and $\mathfrak{p}$ is $\mathfrak{o}_F$ itself.

---

[8]Incidentally, assuming this is the minimal polynomial for y, the rational function $a_0(x)$ is the *norm* of y. As in algebraic number theory, we can use properties of the norm to deduce properties of y. For example, since the total order of zeros and poles of a rational function is 0, the same is true of the total order of zeros and poles of y.

Having observed that the points of $\mathbb{C} \cup \{\infty\}$ correspond to the primes of $\mathbb{C}(x)$, we may expect the points of $\mathcal{S}_F$ to correspond to "primes" of F. This is where ideals come in: *the points of $\mathcal{S}_F$ not at ∞ correspond to the prime ideals of the ring $\mathfrak{o}_F$ of integral algebraic functions of F.*

Namely, for each noninfinity $P \in \mathcal{S}_F$, let

$$\mathfrak{p}_P = \{f \in \mathfrak{o}_F : f(P) = 0\}.$$

(Notice that $\mathfrak{p}_\infty$ is empty because $f(\infty) = 0$ implies $f \notin \mathfrak{o}_F$.) It is clear that $\mathfrak{p}_P$ is an ideal, because it is closed under sums and differences, and under product with arbitrary entire functions. And $\mathfrak{p}_P$ is *maximal*, and hence prime, because if we adjoin to $\mathfrak{p}_P$ any integral algebraic function g not in $\mathfrak{p}_P$ the resulting ideal is the whole $\mathfrak{o}_F$. This is because $g(P) \neq 0$, since $g \notin \mathfrak{p}_P$, but obviously $g - g(P) \in \mathfrak{p}_P$. An ideal containing both g and $g - g(P)$ contains the constant function with value $g(P)$, so it is the whole of $\mathfrak{o}_F$. Thus $\mathfrak{p}_P$ is maximal and hence prime.

Conversely, each prime ideal $\mathfrak{p}$ in O_F equals $\mathfrak{p}_P$ for some point $P \neq \infty$.

To see why, let $\mathfrak{p}$ be a prime ideal, and assign the value 0 to each $f \in \mathfrak{p}$. By the closure properties of ideals, this is a consistent assignment of values to all $f \in \mathfrak{p}$. Since $\mathfrak{p}$ is prime, and hence maximal, we can extend this assignment of values consistently to all of $\mathfrak{o}_F$ by adjoining a constant function with nonzero value c. The assignment then extends consistently to all functions in F by writing them as quotients of integral algebraic functions, and hence it defines a point $P \neq \infty$ (because, as we observed above, functions f with $f(\infty) = 0$ are not integral). Finally, since we assigned the value 0 to all $f \in \mathfrak{p}$, we have

$$\mathfrak{p} = \{f \in \mathfrak{o}_F : f(P) = 0\} = \mathfrak{p}_P.$$

So far, so good. However, the integral algebraic functions and their prime ideals do not tell the whole story of functions on a Riemann surface. The problem lies in the presence of the point ∞ on the Riemann sphere. As we saw in Section 4, the compactness of $\mathbb{C} \cup \{\infty\}$ is crucial in obtaining a nice field of meromorphic functions (the rational function field $\mathbb{C}(x)$). But the presence of ∞ means that the function $w = 1/x$ is just as good as the function x. It is therefore just as good to view the members of $\mathbb{C}(x)$ as functions of $1/x$ as it is to view them as functions of x. So the "integral functions"—the polynomials in x—do not have privileged role in the theory of rational functions.

Indeed, we have already seen that zeros and poles should be treated equally for rational functions, and the same applies to algebraic functions in general. We therefore need a concept that captures the possible zeros and poles of an arbitrary algebraic function, and this is the concept now called a *divisor*.

A divisor $\mathfrak{D}$ on a Riemann surface $\mathcal{S}$ can be written in the (multiplicative) form

$$\mathfrak{D} = P_1^{m_1} P_2^{m_2} \cdots P_k^{m_k},$$

where $P_1, \ldots, P_k$ are points of $\mathcal{S}$ and $m_1, \ldots, m_k$ are nonzero integers. The idea is that if $m_i > 0$ then $P_i^{m_i}$ encodes a *zero of order m_i at P_i* of some function f. In other words, in a neighborhood of P_i, $f(x) = (x - P_i)_i^m E$, where E is approximately constant and nonzero. When $m_1 < 0$, P_i is a similarly defined *pole of order* $-m_i$. For the sake of uniformity, we simply say that P_i is a *point* of order m_i, so points of positive order are zeros and points of negative order are poles.

A divisor is called *principal* (analogous to a principal ideal) if it is realized by some function f, in the sense that f is of order m_i at each point P_i, and of order zero everywhere else. For example we saw in Section 4 that each divisor $P_1^{m_1} P_2^{m_2} \cdots P_k^{m_k}$ on $\mathbb{C} \cup \{\infty\}$ is realized by a rational function, provided that $m_1 + \cdots + m_k = 0$ (and that we need some $P_i = \infty$ if the remaining m_i have a positive sum).

For other Riemann surfaces, as we also know from Section 4, there are further constraints on zeros and poles, and it is more difficult to decide which divisors are principal. We therefore begin by studying a weaker relationship between functions and divisors, called *divisibility* (which helps explain why they are called "divisors"). We say that $P_1^{m_1} P_2^{m_2} \cdots P_k^{m_k}$ *divides* a function f if f has order *at least* m_i at point P_i, and order zero at all other points. It follows, for example, that if P_i divides a product fg then P_i divides f or P_i divides g, so the points behave like prime divisors.

Dedekind and Weber use the word "polygon" instead of "divisor," and they allow only positive m_i. Such divisors are now called "effective" or positive. However, Dedekind and Weber allow quotients of their "polygons," so they essentially use all divisors. Also, they realize a divisor $\mathfrak{D}$ much as they realize an ideal, by a *set* of actual algebraic functions—namely, by the set of all f in the given function field divisible by $\mathfrak{D}$. Then they can define divisibility of divisors exactly as for ideals: "to divide is to contain." It is also not hard to recognize divisibility from the symbols for divisors. For example, P_1 divides P_1^2 and P_1/P_2^2 divides P_1/P_2. In general, a divisor $\mathfrak{a}$ divides divisor $\mathfrak{b}$ if the points in $\mathfrak{b}\mathfrak{a}^{-1}$ occur to powers ≥ 0. With these concepts in place, one can set out on the long road to the theorems of Abel and Riemann-Roch discussed in Sections 1 and 4.

11. From Points to Valuations

The sections §14 and §15 of Dedekind-Weber can be read today as a first introduction to valuation theory—but a general valuation theory in arbitrary fields was first launched in 1913 by Kürschak, prompted by the p-adic ideas of Hensel and without reference to places [that is, points]. The connection between places, valuations, and integral closure present in Dedekind and Weber in the first phase of development was lost in the general valuation theory.

Geyer (1981), p. 120 (my translation).

For each point P of the Riemann surface $\mathcal{S}_F$, Dedekind and Weber define what they call the *order numbers* at P for each function $f \in F$ that either vanishes or becomes infinite at P. Relative to a variable z that is finite and nonzero at P, f has order r at P if $f(z)/z^r$ is nonzero at P but $f(z)/z^{r-1}$ is not. Likewise, $f(z)$ has order $-r$ at P if $f(1/z)$ has order r; that is, if $f(z)$ "goes to infinity like z^r."

Thus each point $P \in \mathcal{S}_F$ defines what we now call a *discrete valuation* on F: a mapping ν_z from F into the integers with the properties that

$$\nu_z(fg) = \nu_z(f) + \nu_z(g),$$
$$\nu_z(f + g) \geq \min(\nu_z(f), \nu_z(g)) \quad \text{if } f + g \neq 0.$$

Since points $P \neq \infty$ correspond to prime ideals $\mathfrak{p}$, the valuation ν_z may also be associated with $\mathfrak{p}$, and indeed

$$\nu_z(f) = \text{exponent of } \mathfrak{p} \text{ in the prime ideal factorization of } f.$$

When viewed in this way, the concept of valuation is unhitched from the existence of "points," and it applies wherever unique prime ideal factorization holds.

Indeed, with hindsight it becomes apparent that a discrete valuation is behind Kummer's concept of "ideal numbers." Kummer does not define his ideal prime numbers $\mathfrak{p}$; in effect, he defines a discrete valuation ν on the cyclotomic integers α such that

$$\nu(\alpha) = \text{exponent of } \mathfrak{p} \text{ in the prime ideal factorization of } \alpha.$$

Even for the field of rational numbers one has an interesting valuation associated with each prime p:

$$\nu(r) = \text{exponent of } p \text{ in the prime factorization of } r.$$

This valuation is called the p-*adic* valuation, and it was introduced by Hensel (1897) to define the p-*adic numbers*, later found to be a powerful tool in algebraic number theory. It seems unlikely that the p-adic valuation would have been considered promising if not for the success of valuations in the theory of algebraic function fields. So here we see influence flowing in the opposite direction: from the theory of function fields back to its origin and model, the theory of algebraic number fields.

In a reversal of the historical development (which is common, but alas seldom mentioned), accounts of the theory of algebraic numbers and functions today often *begin* with the concept of valuation. They then proceed to define "points" (or *places*, as they are now called), ramification, genus, and so on. See for example Artin (1951) or Cohn (1991).

12. Reading the Dedekind-Weber Paper

The influence of Dedekind on the paper is obvious, since it follows his approach to algebraic number theory, almost to a fault. As in Dedekind (1877), the groundwork for ideal theory is laid with a thorough study of modules and their divisibility theory, with the help of concepts such as basis, norm, and discriminant. However, in contrast to Dedekind (1877), the present paper is quite formal, unmotivated, and lacking in historical perspective. This may be because Dedekind was busy working on the third edition of Dirichlet's *Vorlesungen über Zahlentheorie* at the time, so he entrusted the final writeup of the paper to Weber. In the foreword to that edition (dated 11 November 1880) he placed his supplement on ideal theory in a context that included his work with Weber, as follows:

> We mention in particular the extended presentation of ideal theory, in the last supplement, which I first published in the second edition, but in such terse form that the wish for a more detailed version was expressed to me. I have been glad to return to this request since working with my friend H. Weber of Königsberg on an extended investigation, still to appear, whose import is that the same principles may be successfully carried over to the theory of algebraic functions.

In a letter to Weber (30 October 1880, on p. 488 of Dedekind's *Werke*, volume III) Dedekind wrote:

> ... I take this opportunity to express my deepest thanks to you for this work of almost two years, which has given you an infinite amount of trouble, but which has given me the greatest joy and has made a significant scientific advance. It is a beautiful feeling to find a new truth, as Pascal in his first letter to Fermat

expressed so well: Car je voudrais désormais vous ouvrir mon cœur, s'il se pouvrait, tant j'ai de joie de voir notre rencontre. Je vois bien que la vérité est la même à Toulouse et à Paris.[9] I often think back over the progress of our work which, after many oscillations, has acquired more and more the character of inner necessity ...

Another sign of Weber's hand in the paper is that when Weber (1908) returned to the subject in the third volume of his *Algebra*, §§170–201, he gave a very similar treatment—even using the same notation—but with a few extra explanatory remarks.

For whatever reason, the Dedekind-Weber paper is not such easy reading as Dedekind (1877). In an attempt to make the reading easier, I taken the liberty of inserting a short "Summary and Comments" at the beginning of each section (marked by the § symbol). In these insertions I try to summarize the main concepts and results of the section, with comments on where they come from and where they are going. I have sometimes drawn on the very useful commentary of Geyer (1981) in Dedekind et al. (1981), and on the account of the Dedekind and Weber's theory in Koch (1991), Chapter 23. It may also be worthwhile to consult the first chapter of Eichler (1966), which extracts the linear algebraic core of Dedekind and Weber's proof of the Riemann-Roch theorem with remarkable concision. (Eichler's book is also interesting as a modernization of Weber (1908), which is a synthesis of ideas from both Dedekind and Kronecker.)

However, as I remarked in the Preface, all commentators on the Dedekind-Weber paper seem to attend to only some of the ideas it contains, while ignoring others. No doubt my own commentary fails to highlight certain ideas as much as it should, but I hope that the zone of obscurity has at least been reduced, so that any unnoticed gems can be more easily found. Ultimately, of course, the only way to know exactly what Dedekind and Weber are saying is to read their own words.

With only a few exceptions, I have retained Dedekind and Weber's terminology and notation, in order to faithfully represent their train of thought. For example, it is necessary to retain their awkward term "polygon" for what we now call a "divisor." At the time of writing their paper, Dedekind and Weber were prepared to view numbers, functions, ideals, and modules as "divisors," but "polygons" were a new (and more geometric) idea, which they understandably wanted to highlight with a different, and geometric, word.

On the other hand, I think it unnecessarily literal (and tedious) to translate "ganz rational Funktion" as "entire rational function" when "polynomial" better conveys the idea to the modern reader. Similarly, the symbol S for trace ("Spur" in German) has been rewritten as Tr, since this does not clash with the other notation and it carries more meaning today. The only other change in notation, which occurs only in §10, has been to use δ_{rs} for the Kronecker delta in place of Dedekind and Weber's ad hoc notation (r, s). Since the notation $(,)$ now has many meanings (and it already did in the Dedekind-Weber paper) it can only be confusing to retain it when an unambiguous replacement is now available.

---

[9]For I now want to open my heart to you, if I may, because I am so overjoyed by our agreement. I see that the truth is the same in Toulouse as in Paris.

As mentioned in Section 5, Dedekind and Weber use the term *Schaar* for what we now call a vector space, but without establishing the basic vector space properties once and for all. I have decided to translate the word *Schaar* as "vector space," since this seems to be the least confusing option, but readers should be prepared for repeated statements of vector space properties that we now take for granted.

In any case, I have footnoted the places where deviations in terminology or notation occur, so readers are in a position to reconstruct the original if they so desire.

13. Conclusion

The aim of this Introduction has been to expose the historical roots of the theory of algebraic functions, and to give a sense of the mathematical climate at the time when Dedekind and Weber were writing their paper.

Dedekind and Weber (1882) arrived at a time of great ferment in the theory of algebraic functions, with Kronecker (1882) being published in the same journal, and the works of Klein (1882) and Poincaré (1882) published in the same year. Klein and Poincaré championed Riemann's methods and their inspiration from physics—an unpopular view at a time when the goal was to "arithmetize" analysis, under the influence of Weierstrass and Kronecker. The aim of arithmetization was to avoid appeal to geometric and physical intuition by reducing the concepts of analysis to properties of natural numbers and sets of natural numbers. At the time, arithmetization prevailed, and it still does, inasmuch as Kronecker's term "divisor" is now used in all books on Riemann surfaces. However, the analytic approach made something of a comeback after Riemann's ideas were vindicated by Weyl (1913), and today it is likewise true that most books on Riemann surfaces make heavy use of analysis. To see the arithmetic/algebraic approach in its greatest purity, the paper of Dedekind and Weber is probably still the best place to look. The Dedekind-Weber arithmetization was not really the future of analysis, but it *was* the future of algebraic geometry.

John Stillwell

South Melbourne, 1 May 2012

An annotated translation by John Stillwell of

Theorie der algebraischen Functionen einer Veränderlichen.

(Von den Herren *R. Dedekind* in Braunschweig und *H. Weber* in Königsberg.)

Journal für reine und angewandte Mathematik (1882), **92**, 181–290

Introduction

SUMMARY AND COMMENTS

In their introduction, Dedekind and Weber emphasize their two
great innovations: the arithmetic theory of algebraic function fields,
based on Dedekind's theory of algebraic number fields, and the
algebraic definition of Riemann surfaces via "points," each of which
is a consistent assignment of values to the functions of the field.
They do not attempt to extract the common essence of the theories
of function fields and number fields, which was first found by Emmy
Noether in 1927 in the theory of Dedekind rings.

At the end of the introduction they express the hope of return-
ing to the subject "on another occasion," in order to deal with the
question of Abelian integrals and their periods. In fact, they did not
return to the subject, perhaps because such questions are beyond
the scope of their methods. In giving the first rigorous proofs of the
Abel and Riemann-Roch theorems, they had done quite enough.

The purpose of the following investigations is to construct the theory of algebraic functions of one variable, which is one of Riemann's great creations, on
the basis of a simple, yet at the same time rigorous and completely general view-
point. In previous investigations of this topic certain restrictive assumptions about
the singularities of the functions under consideration have been made, as a rule,
and the so-called exceptional cases have either been mentioned casually as limit-
ing cases, or else left aside entirely. Likewise, certain basic theorems on continuity
and developability have been assumed, on the evidence of geometric intuition. A
sounder basis for the fundamental concepts, as well as a general treatment of the
theory, without exceptions, is obtained when one proceeds from a generalization
of the theory of rational functions of one variable, in particular, from the theorem
that each rational function of one variable admits a decomposition into linear factors. This generalization is simple and well known in the first case, in which the
number Riemann denoted by p (called the *genus* by Clebsch) has the value zero.
For the general case, which is related to the one just mentioned in the same way
that general algebraic numbers are related to rationals, the way forward is found
by carrying over to functions those methods that are most successful in the theory
of numbers, including Kummer's creation of ideal numbers.[10]

---

[10]Kummer first introduced ideal numbers in the work *Zur Theorie der complexen Zahlen*
(Crelle's Journal, v. 35). A further development and general presentation of the theory of al-
gebraic numbers may be found in the second and third editions of Dirichlet's *Vorlesungen über
Zahlentheorie*, as well as in the work of Dedekind: *Sur la théorie des nombres entiers algébriques*

If, in analogy with number theory, a *field of algebraic functions* is understood to be a system of such functions with the property that application of the four arithmetic operations to functions of the system always leads to functions of the same system, then this concept coincides with that of the Riemann class of algebraic functions. Any function in such a field can be regarded as the independent variable and the rest as dependent on it. Each such "mode of representation" yields a system of functions of the field called *integral algebraic functions*, the quotients of which exhaust the whole field. Among these integral algebraic functions another group of functions may be distinguished, with the characteristic of polynomials that they admit a greatest common divisor. Such a divisor does not exist in general; however, when theorems on rational functions are related not to the divisor itself, but to the functions divisible by it, they carry over completely to general algebraic functions. In this way one arrives at the concept of *ideal*, a name stemming from Kummer's work in number theory, where nonexistent divisors are replaced by "ideal divisors."

While the present work by no means involves "ideal functions", but only systems of actually existing functions, it seems useful to retain the name "ideal", which is already customary in number theory.

With a suitable definition of multiplication it is possible to calculate with ideals using the same rules as for rational functions. In particular, one has the theorem that each ideal decomposes uniquely into factors that are themselves indecomposable, and hence may be called *prime ideals*. These prime ideals correspond to the linear factors in the theory of polynomials. On this basis one attains a completely precise and general definition of a "point of a Riemann surface", i.e., a complete system of numerical values that can be consistently attached to the functions of the field.

This in turn yields a formal definition of the differential quotient, the genus, and a quite general, elegant presentation of differentials of the first kind. Finally there is the proof of the Riemann-Roch theorem on the number of arbitrary constants in functions defined by their poles, and the theory of differentials of the second and third kinds. Up to this point the continuity and developability of our functions do not come into consideration. It would be completely valid, e.g., to restrict the domain of numbers to algebraic numbers. In this way a well-defined and apparently broader theory of algebraic functions can be treated by methods belonging to its own sphere.

Admittedly all these results come from Riemann's theory using a far narrower range of methods, and as special cases of something more general. However, it is known that a rigorous foundation of Riemann's theory presents certain difficulties and, until these have been completely overcome, it may be that the path we have taken, or at least something similar, is the only one leading to the goal with satisfactory rigor and generality. The theory of ideals itself would be much simpler if one could assume the intuitive concept of a Riemann surface and particularly that of its points, together with the continuity of algebraic functions. Our work, on the other hand, takes a long algebraic detour through the theory of ideals, leading to a completely precise and rigorous definition of a "point of a Riemann surface" that

---

(Paris 1877. Reprinted from the *Bulletin des Sciences math. et astron. of Darboux and Hoüel*). However, knowledge of these publications is not assumed in the present work.

We have learned from verbal communications that for some years Kronecker has been carrying out investigations, related to the work of Weierstrass, based on the same foundations as ours.

can also serve as a basis for the investigation of continuity and related questions. The latter questions, which include those on Abelian integrals and their periods, are left aside for the present. We hope to come back to them on another occasion.

Königsberg, 22 October 1880.

Part I

§1. Fields of algebraic functions

Summary and Comments

In §§1, 2, and 3 of their paper, Dedekind and Weber describe the general framework for their study of algebraic functions: a finite-degree extension of the field $\mathbb{C}(z)$ of rational functions of one variable. The basic concepts are modeled on the analogous concepts in Dedekind's (1877) theory of algebraic integers, where the framework is a finite-degree extension of the field of rational numbers.

Just as an algebraic *number* is the solution of a polynomial equation with rational number coefficients, an algebraic *function* is the solution of a polynomial equation with rational function coefficients. The *degree* of a function is the minimal degree of the equations that it satisfies. As in the number case, there is a proof that sum, difference, product, and quotient of algebraic functions are algebraic. Indeed, if θ is a function of degree n, then the linear combinations of $1, \theta, \ldots, \theta^{n-1}$ with rational function coefficients form a *field of degree* n, with basis $\{1, \theta, \ldots, \theta^{n-1}\}$. Such *algebraic function fields* Ω are the basic object of study in what follows.

A variable θ is called an algebraic function of an independent variable z when it satisfies an irreducible algebraic equation

(1)
$$F(\theta, z) = 0.$$

Here F denotes an expression of the form

$$F(\theta, z) = a_0\theta^n + a_1\theta^{n-1} + \cdots + a_{n-1}\theta + a_n,$$

where the coefficients $a_0, a_1, \ldots, a_n$ are polynomial functions of z without common divisor. The assumed irreducibility of equation (1) means that θ does not satisfy an equation of lower degree in θ and, as follows from the algorithm for greatest common divisor, if

$$G(\theta, z) = b_0\theta^m + b_1\theta^{m-1} + \cdots + b_{m-1}\theta + b_m = 0$$

is a second equation satisfied by θ then $G(\theta, z)$ must be divisible by $F(\theta, z)$. It may now be proved that $G(\theta, z)$ also cannot be of lower degree in z than $F(\theta, z)$, and it is of the same degree only if $G(\theta, z)$ differs from $F(\theta, z)$ by a factor independent of z. If we assume that the coefficients $b_0, b_1, \ldots, b_m$ are freed of common factors, and if we let

$$H(\theta, z) = c_0\theta^{m-n} + c_1\theta^{m-n-1} + \cdots + c_{m-n}$$

45

denote the denominator-free quotient of G by F, then

$$kG(\theta, z) = F(\theta, z) \cdot H(\theta, z),$$

where k is a polynomial function of z. Comparison of coefficients yields

$$kb_0 = a_0 c_0,$$
$$kb_1 = a_0 c_1 + a_1 c_0,$$
$$kb_2 = a_0 c_2 + a_1 c_1 + a_2 c_0,$$
$$\dots\dots\dots\dots\dots\dots\dots\dots$$

where the $c_0, c_1, \ldots, c_{m-n}$ can likewise be assumed to have no common divisor.

It follows from this, first, that k must be constant, so that it can be set equal to 1. Because if $a_0, a_1, \ldots, a_{r-1}, c_0, c_1, \ldots, c_{s-1}$ are divisible by a linear factor of k and a_r, c_s are not, then the equation

$$kb_{r+s} = \cdots + a_{r-1} c_{s+1} + a_r c_s + a_{r+1} c_{s-1} + \cdots$$

yields the contradiction that $a_r c_s$ is divisible by the same linear factor. It then follows that the degree of $G(\theta, z)$ in z is the sum of the degrees of F and H in z, because if a_r, c_s are the first among the coefficients a, c to have degrees of maximal value then it again follows from

$$b_{r+s} = \cdots + a_{r-1} c_{s+1} + a_r c_s + a_{r+1} c_{s-1} + \cdots$$

that the degree of b_{r+s} equals the sum of the degrees of a_r and c_s.

If one divides equation (1) by a_0 then it can also be written in the form

$$(2) \qquad f(\theta, z) = \theta^n + b_1 \theta^{n-1} + \cdots + b_{n-1} \theta + b_n = 0,$$

where the coefficients $b_1, b_2, \ldots, b_n$ may be fractional rational functions of z.

The system $\Phi(\theta, z)$ of all rational functions of θ and z has the property of closure under the elementary arithmetic operations of addition, subtraction, multiplication and division, and this system is therefore called a *field Ω of algebraic functions of degree n.* If $\varphi(\theta)$ is a *polynomial* function of θ whose coefficients are rational in z, then by algebraic division one can determine two functions $q(\theta), r(\theta)$, the second of which with degree no greater than $n - 1$, such that

$$\varphi(\theta) = q(\theta) f(\theta) + r(\theta).$$

Then, by (2)

$$\varphi(\theta) = r(\theta).$$

If $\varphi(\theta)$ is not divisible by $f(\theta)$ then (by the assumed irreducibility of $f(\theta)$) these two functions have no common divisor and hence, by the method of finding the greatest common divisor, one obtains two functions $f_1(\theta), \varphi_1(\theta)$ such that

$$f(\theta) f_1(\theta) + \varphi(\theta) \varphi_1(\theta) = 1,$$

and hence by (2)

$$\varphi_1(\theta) = \frac{1}{\varphi(\theta)}.$$

These remarks, along with the assumed irreducibility of $f(\theta)$, yield the following:

Theorem. *Each function ζ in the field Ω is uniquely expressible in the form*

$$\zeta = x_0 + x_1 \theta + \cdots + x_{n-1} \theta^{n-1},$$

where the coefficients $x_0, x_1, \ldots, x_{n-1}$ are rational functions of z. Conversely, each function of this form obviously belongs to the field Ω.

If one chooses n functions arbitrarily from the field Ω:

$$\eta_1 = x_0^{(1)} + x_1^{(1)}\theta + \cdots + x_{n-1}^{(1)}\theta^{n-1},$$
$$\eta_2 = x_0^{(2)} + x_1^{(2)}\theta + \cdots + x_{n-1}^{(2)}\theta^{n-1},$$
$$\cdots\cdots\cdots\cdots\cdots\cdots\cdots\cdots\cdots\cdots\cdots$$
$$\eta_n = x_0^{(n)} + x_1^{(n)}\theta + \cdots + x_{n-1}^{(n)}\theta^{n-1},$$

but in such a way that the determinant[11]

$$\sum \pm x_0^{(1)} x_1^{(2)} \cdots x_{n-1}^{(n)}$$

is not identically zero, then it follows that every function in the field Ω is expressible in the form

$$\zeta = y_1\eta_1 + y_2\eta_2 + \cdots + y_n\eta_n,$$

where the coefficients $y_1, y_2, \ldots, y_n$ are rational functions of z. Such a system of functions $\eta_1, \eta_2, \ldots, \eta_n$ will be called a *basis of the field* Ω.

A system of functions $\eta_1, \eta_2, \ldots, \eta_n$ in the field Ω is a basis if and only if they satisfy no equation (identity) of the form

$$y_1\eta_1 + y_2\eta_2 + \cdots + y_n\eta_n = 0$$

in which the coefficients $y_1, y_2, \ldots, y_n$ do not all vanish. For example, the functions $1, \theta, \theta^2, \ldots, \theta^{n-1}$ form a basis of Ω.

§2. Norm, trace, and discriminant

Summary and Comments

Just as every algebraic number α in a field of finite degree has a norm $N(\alpha)$, which is a rational number, so does each algebraic function ζ in a field Ω of degree n have a norm $N(\zeta)$, which is a rational function. The basic properties of the algebraic function norm are like those of the algebraic number norm: the norm of a rational function is its nth power, and the norm is *multiplicative*: $N(\zeta\zeta') = N(\zeta)N(\zeta')$. As with algebraic numbers, this often allows divisibility questions for algebraic objects to be reduced to divisibility questions for rational objects.

But in this section, the norm comes from a determinant—the characteristic polynomial of multiplication by ζ on Ω—rather than as a product of conjugates, as in the algebraic number case. The product of conjugates is somewhat more subtle in the function case, and it is taken up much later, in §15.

The present approach has the advantage of revealing another useful concept: the *trace* $Tr(\zeta)$. $N(\zeta)$ and $Tr(\zeta)$ are, up to sign, the coefficients of ζ^0 and ζ^{n-1} in the characteristic polynomial equation satisfied by ζ. The basic property of the trace is that it is *additive*:

$$Tr(\zeta + \zeta') = Tr(\zeta) + Tr(\zeta').$$

---

[11]The curious shorthand $\sum \pm a_{11}a_{22}\cdots a_{nn}$ for the determinant with (i,j)-entry a_{ij} is often used by Dedekind and Weber in this paper. (Translator's note.)

The third important concept in this section, the *discriminant*, is likewise motivated by the concept of the same name in Dedekind's (1877) algebraic number theory. The discriminant $\Delta(\eta_1, \eta_2, \ldots, \eta_n)$ of functions $\eta_1, \eta_2, \ldots, \eta_n \in \Omega$ is a determinant that tests whether $\eta_1, \eta_2, \ldots, \eta_n$ form a basis of Ω, being nonzero if and only if they do. The vanishing of $\Delta(\eta_1, \eta_2, \ldots, \eta_n)$ when $\eta_1, \eta_2, \ldots, \eta_n$ do *not* form a basis of Ω also follows from the *fundamental theorem on discriminants*, namely

$$\Delta(\eta_1', \eta_2', \ldots, \eta_n') = X^2 \Delta(\eta_1, \eta_2, \ldots, \eta_n),$$

where X is the determinant of the linear equations expressing the functions $\eta_1', \eta_2', \ldots, \eta_n'$ in terms of the basis elements $\eta_1, \eta_2, \ldots, \eta_n$.

Given an arbitrary basis $\eta_1, \eta_2, \ldots, \eta_n$ for the representation of functions in Ω, and any function ζ, one can set

(1)
$$\begin{cases} \zeta\eta_1 = y_{1,1}\eta_1 + y_{1,2}\eta_2 + \cdots + y_{1,n}\eta_n, \\ \zeta\eta_2 = y_{2,1}\eta_1 + y_{2,2}\eta_2 + \cdots + y_{2,n}\eta_n, \\ \cdots\cdots\cdots\cdots\cdots\cdots\cdots\cdots\cdots\cdots \\ \zeta\eta_n = y_{n,1}\eta_1 + y_{n,2}\eta_2 + \cdots + y_{n,n}\eta_n, \end{cases}$$

where the coefficients $y_{\iota,\iota'}$ are rational functions of z. This yields the equation

(2)
$$\begin{vmatrix} y_{1,1} - \zeta & y_{1,2} & \cdots & y_{1,n} \\ y_{2,1} & y_{2,2} - \zeta & \cdots & y_{2,n} \\ \cdots & \cdots & \cdots & \cdots \\ y_{n,1} & y_{n,2} & \cdots & y_{n,n} - \zeta \end{vmatrix} = 0,$$

which, when arranged according to powers of ζ, has the form

(3)
$$\varphi(\zeta) = \zeta^n + b_1\zeta^{n-1} + \cdots + b_{n-1}\zeta + b_n = 0.$$

And it follows easily from the multiplication theorem for determinants[12] that this equation is independent of the choice of basis $\eta_1, \eta_2, \ldots, \eta_n$. Among the coefficients $b_1, b_2, \ldots, b_n$ of the function φ, which are all rational functions of z and completely determined by ζ, there are two that are given special names, due to their importance in what follows. The function

(4)
$$(-1)^n b_n = \begin{vmatrix} y_{1,1} & y_{1,2} & \cdots & y_{1,n} \\ y_{2,1} & y_{2,2} & \cdots & y_{2,n} \\ \cdots & \cdots & \cdots & \cdots \\ y_{n,1} & y_{n,2} & \cdots & y_{n,n} \end{vmatrix}$$

is called the *norm* of the function ζ and is denoted by $N(\zeta)$. This function is subject to the following theorems.

1. The only function whose norm vanishes identically is the "zero" function. Because if $N(\zeta) = 0$ in system (1) it follows that there is a system $y_1, y_2, \ldots, y_n$ of rational functions of z, not all vanishing, such that

$$\zeta(y_1\eta_1 + y_2\eta_2 + \cdots + y_n\eta_n) = 0.$$

Then, since $\eta_1, \eta_2, \ldots, \eta_n$ is a basis of Ω, $\zeta = 0$.

---

[12]That is, the theorem that the determinant is multiplicative. (Translator's note.)

2. The norm of a rational function of z is the nth power of the function. Because if ζ is rational the equations (1) reduce to the identities $\zeta \eta_h = \zeta \eta_h$, whence $N(\zeta) = \zeta^n$.

3. If ζ' is a second function in the field Ω and if the equation system for this function, corresponding to (1), is

$$\zeta' \eta_h = \sum_\iota y'_{h,\iota} \eta_\iota,$$

then it follows that

$$\zeta \zeta' \eta_h = \sum_{\iota,\iota'} y_{h,\iota} y'_{\iota,\iota'} \eta_{\iota'},$$

whence

$$N(\zeta\zeta') = N(\zeta)N(\zeta')$$

by the multiplication theorem for determinants.

4. It follows from 2 and 3 that

$$N(\zeta)N\left(\frac{1}{\zeta}\right) = 1,$$

hence

$$N\left(\frac{\zeta}{\zeta'}\right) = \frac{N(\zeta)}{N(\zeta')}.$$

5. Finally, the definition of the function φ, together with (2) and (3), yields the important theorem: if t is an arbitrary constant (or a rational function of z) then

$$\varphi(t) = N(t - \zeta).$$

We now call the function

(5) $-b_1 = y_{1,1} + y_{2,2} + \cdots + y_{n,n}$

the *trace* of ζ and denote it by $Tr(\zeta)$.[13] The following theorems about it follow immediately from the definition:

(6) $Tr(0) = 0,$

(7) $Tr(1) = n.$

Also, when x is a rational function of z and ζ, ζ' are two functions in Ω:

(8) $Tr(x\zeta) = xTr(\zeta),$

(9) $Tr(\zeta + \zeta') = Tr(\zeta) + Tr(\zeta').$

It follows from these considerations that *every function ζ in Ω satisfies an nth degree equation $\varphi(\zeta) = 0$ whose coefficients are rational in z*. When this equation is irreducible, the functions $1, \zeta, \zeta^2, \ldots, \zeta^{n-1}$ form a basis of Ω. If not, let

(10) $\varphi_1(\zeta) = \zeta^e + b'_0 \zeta^{e-1} + \cdots + b'_{e-1}\zeta + b'_e = 0$

be the equation of lowest degree, with rational coefficients in z, satisfied by ζ. Thus $\varphi_1(\zeta) = 0$ is irreducible and $e < n$. Since $\varphi(\zeta)$ also vanishes, $\varphi(\zeta)$ must be

---

[13]Dedekind denotes it by $S(\zeta)$ because the German word is "Spur." (Translator's note.)

algebraically divisible by $\varphi_1(\zeta)$, and it follows as in §1 that each rational function η of z and ζ is representable in the form

$$\eta = x_0 + x_1\zeta + \cdots + x_{e-1}\zeta^{e-1},$$

where the coefficients $x_0, x_1, \ldots, x_{e-1}$ are rational[14] in z. Now if

$$\theta^f + \eta_1\theta^{f-1} + \cdots + \eta_{f-1}\theta + \eta_f = 0$$

is the equation of lowest degree satisfied by θ with coefficients rational in z *and* ζ, then the $e \cdot f$ functions

(11) $\zeta^h\theta^k$ $(h = 0, 1, \ldots, e - 1; k = 0, 1, \ldots, f - 1)$

satisfy no linear equation with coefficients rational in z, while *each* function in Ω is representable as a *linear* combination of these functions with coefficients rational in z. It follows that the functions (11) form a basis of Ω, so that

$$e \cdot f = n$$

and e is a divisor of n.

If one uses the basis (11) to construct the norm of ζ, then one sees easily from equation (10) that

$$N(\zeta) = ((-1)^e b_e')^f = (-1)^n b_e'^f.$$

Since, in addition, the function $\zeta - t$ for any constant t satisfies an equation of the same degree as ζ does, we have the theorem:

6.[15]   *The function $\varphi(t)$ in (3) is either irreducible or an integer power of an irreducible function.*

If $\eta_1, \eta_2, \ldots, \eta_n$ is any system of n functions in Ω, whether a basis or not, then we associate with this system a rational function of z called the *discriminant*, denoted by $\Delta(\eta_1, \eta_2, \ldots, \eta_n)$ and defined as follows

(12) $$\Delta(\eta_1, \eta_2, \ldots, \eta_n) = \begin{vmatrix} Tr(\eta_1\eta_1) & Tr(\eta_1\eta_2) & \ldots & Tr(\eta_1\eta_n) \\ Tr(\eta_2\eta_1) & Tr(\eta_2\eta_2) & \ldots & Tr(\eta_2\eta_n) \\ \ldots & \ldots & \ldots & \ldots \\ Tr(\eta_n\eta_1) & Tr(\eta_n\eta_2) & \ldots & Tr(\eta_n\eta_n) \end{vmatrix}.$$

The discriminant is identically zero if and only if the functions $\eta_1, \eta_2, \ldots, \eta_n$ form a basis of Ω.

To prove the first part of this assertion, we suppose that $\Delta(\eta_1, \eta_2, \ldots, \eta_n) = 0$. Under this hypothesis there is a system of rational functions $y_1, y_2, \ldots, y_n$ of z, which do not all vanish, such that

$$y_1 Tr(\eta_1\eta_k) + y_2 Tr(\eta_2\eta_k) + \cdots + y_n Tr(\eta_n\eta_k) = Tr(\eta_k(y_1\eta_1 + y_2\eta_2 + \cdots + y_n\eta_n))$$

$$= 0 \qquad\qquad (k = 1, 2, \ldots, n).$$

---

[14]The equation $\varphi_1(\zeta) = 0$ gives rise to a field Ω_1 of degree e, of algebraic functions that also belong to Ω, and hence it may be called a divisor of the field Ω. (Translator's note: This use of the word "divisor" to denote a substructure is at odds with the arithmetical definition, as Dedekind and Weber concede when they define divisibility of modules in §4, 7.)

[15]This subsection is wrongly labelled 10 in the original. However, no confusion arises, because it is never cross referenced. (Translator's note.)

Hence if one chooses an arbitrary system $x_1, x_2, \ldots, x_n$ of rational functions of z and sets

$$y_1\eta_1 + y_2\eta_2 + \cdots + y_n\eta_n = \eta,$$
$$x_1\eta_1 + x_2\eta_2 + \cdots + x_n\eta_n = \xi,$$

then it follows that

$$Tr(\xi\eta) = 0.$$

But when the functions $\eta_1, \eta_2, \ldots, \eta_n$ form a basis of Ω the function ξ can be arbitrary in Ω, and since η does not vanish, it can for example be $\frac{1}{\eta}$. The last equation is then certainly *not* satisfied, and hence under this hypothesis the discriminant of $\eta_1, \eta_2, \ldots, \eta_n$ does *not* vanish identically.

If we now assume that $\eta_1, \eta_2, \ldots, \eta_n$ form a basis of Ω and set

$$\eta_k' = x_{1,k}\eta_1 + x_{2,k}\eta_2 + \cdots + x_{n,k}\eta_n, \qquad (k = 1, 2, \ldots, n)$$

then the functions $\eta_k', \eta_2', \ldots, \eta_n'$ form a basis of Ω, or not, according as the determinant

$$X = \sum \pm x_{1,1}x_{2,2}\cdots x_{n,n}$$

of the rational functions $x_{h,k}$ of z is nonzero or not. But

$$Tr(\eta_h'\eta_k') = \sum_{1 \le \iota, \iota' \le n} x_{\iota,h}x_{\iota',k}Tr(\eta_\iota\eta_{\iota'}),$$

and this, together with the multiplication theorem for determinants, yields the *fundamental theorem on discriminants*

(13) $$\Delta(\eta_1', \eta_2', \ldots, \eta_n') = X^2\Delta(\eta_1, \eta_2, \ldots, \eta_n),$$

which also shows the correctness of the second part of the assertion above, namely, that the discriminant of a system of functions vanishes when they do not form a basis for Ω.

§3. The system of integral algebraic functions of z in the field Ω

SUMMARY AND COMMENTS

Having established a suitable framework for algebraic function theory—a finite-degree extension Ω of the field of rational functions—Dedekind and Weber now identify the "integers" of Ω: the *integral* algebraic functions. These functions form a ring and have properties analogous to those of the algebraic integers studied by Dedekind (1877). For example, just as the norm of an algebraic integer is an ordinary integer, the norm of an integral algebraic function is an "integral rational function," that is, a polynomial.

One finds throughout that polynomials appear in the theory of algebraic functions where ordinary integers appear in the theory of algebraic integers.

The appropriate concept of basis for integral algebraic functions, naturally, is an independent collection of such functions, the linear combinations of which (with polynomial coefficients) are all the integral algebraic functions in Ω. It is shown that such "integral bases" exist, and that they all have the same discriminant. This invariant discriminant is called the *discriminant of* Ω.

Definition. A function ω in the field Ω is called an *integral algebraic* function of z when, in the equation of lowest degree satisfied by ω according to §2,

$$(1) \qquad \varphi(\omega) = \omega^e + b_1\omega^{e-1} + \cdots + b_{e-1}\omega + b_e = 0,$$

the coefficients $b_1, b_2, \ldots, b_e$ are *polynomial* functions of z. In the contrary case it is called a *fractional* function. The collection of all integral algebraic functions of z in Ω will be denoted by $\mathfrak{o}$. Since, by §2, $N(t - \omega)$ is an integer power of $\varphi(t)$, it follows that if ω is an integral algebraic function then all the coefficients of $N(t-\omega)$ are polynomial functions of z, hence in particular:

1. The norm and trace of integral algebraic functions are polynomial functions of z.

In addition, the definition of integral algebraic function yields:

2. A rational function of z belongs to the system $\mathfrak{o}$ if and only if it is a polynomial function of z.

3. Each function η in Ω can be converted into a function in $\mathfrak{o}$ by multiplication by a nonzero polynomial function of z. By §2, η satisfies an equation of minimal degree of the form

$$b_0\eta^e + b_1\eta^{e-1} + \cdots + b_{e-1}\eta + b_e = 0,$$

whose coefficients are polynomial functions of z, and this becomes an equation of the form (1) for ω under the substitution $b_0\eta = \omega$.

4. A function ω in the field Ω that satisfies any equation of the form

$$\psi(\omega) = \omega^m + c_1\omega^{m-1} + \cdots + c_{m-1}\omega + c_m = 0,$$

in which the coefficients $c_1, \ldots, c_m$ are polynomial functions of z, is an integral algebraic function. Because if

$$\varphi(\omega) = \omega^e + b_1\omega^{e-1} + \cdots + b_{e-1}\omega + b_e = 0$$

is the equation of lowest degree satisfied by ω then $\psi(\omega)$ must be algebraically divisible by $\varphi(\omega)$:

$$\psi(\omega) = \varphi(\omega)\chi(\omega),$$

and it is easily seen that this implies that the coefficients of $\varphi(\omega)$ and $\chi(\omega)$ are also polynomial functions of z (Gauss, *Disquisitiones Arithmeticae*, art. 42).[16] This yields the fundamental theorem on integral algebraic functions:

5. *The sum, difference and product of two integral algebraic functions are also integral algebraic functions.*

---

[16]This is a reference to the famous *Gauss's lemma*, which is actually about polynomials with rational *number* coefficients, but a similar lemma holds for polynomials with rational function coefficients. (Translator's note.)

Namely, if ω', ω'' are two integral algebraic functions in Ω that satisfy the respective equations

$$\omega'^{n'} + b_1' \omega'^{n'-1} + \cdots + b_{n'-1}' \omega' + b_{n'}' = 0,$$

$$\omega''^{n''} + b_1'' \omega''^{n''-1} + \cdots + b_{n''-1}'' \omega'' + b_{n''}'' = 0,$$

then one can, by letting $\omega_1, \omega_2, \ldots, \omega_m$ be the $m = n'n''$ products

$$\omega'^{h'} \omega''^{h''} \qquad (h' = 0, 1, \ldots, n'-1; h'' = 0, 1, \ldots, n''-1)$$

and letting ω denote one of the three functions $\omega' \pm \omega'', \omega'\omega''$, set:[17]

$$\omega\omega_1 = x_{1,1}\omega_1 + \cdots + x_{1,m}\omega_m,$$

$$\cdots\cdots\cdots\cdots\cdots\cdots\cdots\cdots\cdots\cdots\cdots$$

$$\omega\omega_m = x_{m,1}\omega_1 + \cdots + x_{m,m}\omega_m,$$

where the $x_{h,h'}$ are polynomial functions of z. And from this one obtains

$$\begin{vmatrix} x_{1,1} - \omega & x_{1,2} & \cdots & x_{1,m} \\ x_{2,1} & x_{2,2} - \omega & \cdots & x_{2,m} \\ \cdots & \cdots & \cdots & \cdots \\ x_{m,1} & x_{m,2} & \cdots & x_{m,m} - \omega \end{vmatrix} = 0,$$

which is an equation for ω whose coefficients are polynomial functions of z.

This yields the corollary that each polynomial function of functions in $\mathfrak{o}$ is itself in $\mathfrak{o}$.

6. An integral algebraic function ω is said to be *divisible* by another integral algebraic function ω' when there is a third integral algebraic function ω'' satisfying the condition

$$\omega = \omega'\omega''.$$

It follows immediately from this definition that:

If ω is divisible by ω', and ω' by ω'', then ω is also divisible by ω''.

If ω' and ω'' are divisible by ω then $\omega' \pm \omega''$ is also divisible by ω, and in general: if $\omega_1, \omega_2, \omega_3, \ldots$ are divisible by ω and if $\omega_1', \omega_2', \omega_3', \ldots$ are any functions in $\mathfrak{o}$, then $\omega_1'\omega_1 + \omega_2'\omega_2 + \omega_3'\omega_3 + \cdots$ is also divisible by ω.

7. If the functions $\eta_1, \eta_2, \ldots, \eta_n$ form a basis of Ω, then (by 3) one can determine n nonzero polynomial functions $a_1, a_2, \ldots, a_n$ of z so that

$$\omega_1 = a_1\eta_1, \quad \omega_2 = a_2\eta_2, \quad \ldots, \quad \omega_n = a_n\eta_n$$

are integral algebraic functions, and the latter likewise form a basis of Ω, since

$$\Delta(\omega_1, \omega_2, \ldots, \omega_n) = a_1^2 a_2^2 \cdots a_n^2 \Delta(\eta_1, \eta_2, \ldots, \eta_n)$$

is nonzero. Thus Ω has bases $\omega_1, \omega_2, \ldots, \omega_n$ consisting wholly of integral algebraic functions, and the discriminant of such a basis is itself a nonzero polynomial function of z, since the $Tr(\omega_r\omega_s)$ are polynomial functions of z. Each function of the form

(2) $$\omega = x_1\omega_1 + x_2\omega_2 + \cdots + x_n\omega_n,$$

---

[17]This is essentially the same as the argument by which Dedekind (1871), p. 437, proves that the algebraic integers are closed under sum, difference and product. (Translator's note.)

in which the $x_1, x_2, \ldots, x_n$ are polynomial functions of z, then belongs to the system $\mathfrak{o}$. However, it certainly does not follow that, conversely, each function in $\mathfrak{o}$ is representable in this form.

If we suppose that $\mathfrak{o}$ includes functions not of the form (2) then it is possible to choose a linear function $z - c$ and certain polynomial functions $x_1, x_2, \ldots, x_n$, not all divisible by $z - c$, so that

$$\frac{x_1\omega_1 + x_2\omega_2 + \cdots + x_n\omega_n}{z - c}$$

is an integral algebraic function. The functions $x_1, x_2, \ldots, x_n$ may now be reduced to their constant, and not all zero, remainders $c_1, c_2, \ldots, c_n$ modulo $z - c$, and one sees that

$$\omega = \frac{c_1\omega_1 + c_2\omega_2 + \cdots + c_n\omega_n}{z - c}$$

is also an integral algebraic function. If c_1 is nonzero, then the n integral algebraic functions

$$\omega \quad \text{and} \quad \omega_2, \omega_3, \ldots, \omega_n$$

also form a basis of Ω and at the same time, by §2 (13),

$$\Delta(\omega, \omega_2, \ldots, \omega_n) = \frac{c_1^2}{(z - c)^2} \Delta(\omega_1, \omega_2, \ldots, \omega_n)$$

is of lower degree than $\Delta(\omega_1, \omega_2, \ldots, \omega_n)$. Now, since both these discriminants are polynomial functions of z, by repeating this process one finally arrives at a basis of Ω consisting of integral algebraic functions $\omega_1', \omega_2', \ldots, \omega_n'$ with discriminant of a degree that cannot be lowered, and consequently *each function ω in $\mathfrak{o}$ is of the form*

$$\omega = x_1\omega_1' + x_2\omega_2' + \cdots + x_n\omega_n'$$

with polynomial functions of z as coefficients. Such a system will be called a *basis of $\mathfrak{o}$.*

If $\omega_1, \omega_2, \ldots, \omega_n$ is a basis of $\mathfrak{o}$ and if

$$\omega_\iota' = x_{\iota,1}\omega_1 + x_{\iota,2}\omega_2 + \cdots + x_{\iota,n}\omega_n, \qquad (\iota = 1, 2, \ldots, n)$$

then the system $\omega_1', \omega_2', \ldots, \omega_n'$ is a basis of $\mathfrak{o}$ if and only if the determinant

$$X = \sum \pm x_{1,1}x_{2,2}\cdots x_{n,n}$$

of the polynomial functions $x_{\iota,\iota'}$ is a nonzero *constant.* Because if this determinant has any linear factor $z - c$ then it is possible to determine constants $c_1, c_2, \ldots, c_n$, not all zero, so that the n polynomials in z

$$c_1 x_{1,\iota} + c_2 x_{2,\iota} + \cdots + c_n x_{n,\iota}$$

are divisible by $z - c$ (i.e., they vanish for $z = c$). But then

$$\frac{c_1\omega_1' + c_2\omega_2' + \cdots + c_n\omega_n'}{z - c}$$

is an integral algebraic function and hence $\omega_1', \omega_2', \ldots, \omega_n'$ is not a basis of $\mathfrak{o}$.

On the other hand, since

$$\Delta(\omega_1', \omega_2', \ldots, \omega_n') = X^2 \Delta(\omega_1, \omega_2, \ldots, \omega_n),$$

it follows that the discriminant of a basis of $\mathfrak{o}$ is independent of the choice of basis, apart from a constant factor. Thus one obtains a definite polynomial in z when one divides the discriminant of any basis of $\mathfrak{o}$ by the coefficient of the highest power of

z. The latter function will be called the *discriminant of the field* Ω *or of the system* $\mathfrak{o}$, and it will be denoted by $\Delta(\Omega)$ or $\Delta(\mathfrak{o})$.

§4. Modules of functions

SUMMARY AND COMMENTS

To pave the way for the study of ideals of integral algebraic functions, Dedekind and Weber now introduce the concept of a *module*: a set of algebraic functions closed under the operations of addition, subtraction, and multiplication by polynomials. The analogous concept of module in algebraic number theory, studied in §1 of Dedekind (1877), requires only closure under addition and subtraction (but of course these yield closure under multiplication by integers). The initial development of the module concept is the same in both settings. As usual, there is a concept of basis, and all bases of a module have the same size.

What is special, and crucial, about modules is the concept of *divisibility*. It is motivated by the concept of divisibility for ideals: "to divide is to contain." That is, a module $\mathfrak{b}$ *divides* module $\mathfrak{a}$ if $\mathfrak{b}$ *contains* $\mathfrak{a}$. In conformity with this concept of divisibility, the *least common multiple* of two modules $\mathfrak{a}, \mathfrak{b}$ is simply their intersection $\mathfrak{a} \cap \mathfrak{b}$. Not so obviously, the *greatest common divisor* of $\mathfrak{a}, \mathfrak{b}$ is the set of sums $\alpha + \beta$, where $\alpha \in \mathfrak{a}$ and $\beta \in \mathfrak{b}$.

The *product* $\mathfrak{ab}$ is defined to be the set of all finite sums $\sum \alpha\beta$, where $\alpha \in \mathfrak{a}$ and $\beta \in \mathfrak{b}$. As Dedekind and Weber point out, it is *not* generally the case that $\mathfrak{a}$ divides $\mathfrak{ab}$. However, this turns out to be the case for ideals (see §9, 4).

In what follows we consider systems of functions that we call *modules of functions* or simply *modules*, and define as follows. *A function system (in Ω) is called a module when it is closed under addition, subtraction and multiplication by polynomial functions of z.*

If $\alpha_1, \alpha_2, \ldots, \alpha_m$ are m given functions and $x_1, x_2, \ldots, x_m$ are arbitrary polynomial functions of z, then the collection of all functions of the form

$$\alpha = x_1\alpha_1 + x_2\alpha_2 + \cdots + x_m\alpha_m$$

is a module. Such a module is said to be *finitely generated*[18] and is denoted by

$$\mathfrak{a} = [\alpha_1, \alpha_2, \ldots, \alpha_m].$$

The system of functions $\alpha_1, \alpha_2, \ldots, \alpha_m$ is called a *basis* of this module.

We call a function system $\alpha_1, \alpha_2, \ldots, \alpha_m$ *rationally irreducible*, or call the functions $\alpha_1, \alpha_2, \ldots, \alpha_m$ *rationally independent*, when an equation of the form

$$x_1\alpha_1 + x_2\alpha_2 + \cdots + x_m\alpha_m = 0$$

---

[18]In the language typically used for finitely generated structures at that time, Dedekind and Weber call the module "finite," rather than "finitely generated." Also, their modules are always *sub*modules, of the function field in this case, and they omit mention of the ring the module is over, in this case the polynomial ring $\mathbb{C}[z]$. (Translator's note.)

holds for rational x only if $x_1 = 0$, $x_2 = 0$, ..., $x_m = 0$. A basis of the field Ω is therefore rationally irreducible, and there is no system of more than n rationally independent functions in Ω.

We now prove the theorem:

1. *Each finitely generated module has a rationally independent basis.*

The proof is obtained immediately from the following lemma:

If the polynomial functions $y_{1,1}, y_{2,1}, \ldots, y_{m,1}$ have no common divisor, then one may determine other polynomial functions $y_{1,2}, y_{2,2}, \ldots, y_{m,m}$ so that

$$\sum \pm y_{1,1} y_{2,2} \cdots y_{m,m} = 1.^{19}$$

Now if the functions $\alpha_1, \alpha_2, \ldots, \alpha_m$ satisfy an equation

$$\sum_{\iota=1}^{m} y_{\iota,1} \alpha_\iota = 0,$$

in which the polynomials $y_{1,1}, \ldots, y_{m,1}$ can be assumed to have no common divisor, one sets

$$\sum_{\iota=1}^{m} y_{\iota,2} \alpha_\iota = \beta_2,$$

$$\cdots\cdots\cdots\cdots\cdots$$

$$\sum_{\iota=1}^{m} y_{\iota,m} \alpha_\iota = \beta_m.$$

Then the module $[\alpha_1, \alpha_2, \ldots, \alpha_m]$ is identical with the module $[\beta_1, \beta_2, \ldots, \beta_m]$ whose basis contains one function less. If the functions β_ι are still not rationally independent then one can reduce further in the same way, and finally arrive at an irreducible basis except in the case where the functions α_ι all vanish (a case we exclude from the concept of module). From now on we shall always understand a basis to be an irreducible basis.

2. Although the preceding process finds many different bases for the same module, the number of functions in the basis is always the same, since a system with more functions cannot be rationally irreducible. Then if $\alpha_1, \alpha_2, \ldots, \alpha_m$ and

---

[19]The result is correct and well known for $m = 2$. Thus if we suppose it proved for $m - 1$ we can satisfy the equation

$$\begin{vmatrix} y_{1,1} & y_{2,1} & \cdots & y_{m-1,1} \\ y_{1,3} & y_{2,3} & \cdots & y_{m-1,3} \\ \cdots & \cdots & \cdots & \cdots \\ y_{1,m} & y_{2,m} & \cdots & y_{m_1,m} \end{vmatrix} = \partial,$$

where ∂ denotes the greatest common divisor of $y_{1,1}, y_{2,1}, \ldots, y_{m-1,1}$, and hence when we determine polynomials x, y so that $x y_{m,1} - y \partial = (-1)^{m-1}$ it follows that

$$\begin{vmatrix} y_{1,1} & y_{2,1} & \cdots & y_{m-1,1} & y_{m,1} \\ \frac{x y_{1,1}}{\partial} & \frac{x y_{2,1}}{\partial} & \cdots & \frac{x y_{m-1,1}}{\partial} & y \\ y_{1,3} & y_{2,3} & \cdots & y_{m-1,3} & 0 \\ \cdots & \cdots & \cdots & \cdots & \cdots \\ y_{1,m} & y_{2,m} & \cdots & y_{m-1,m} & 0 \end{vmatrix} = 1.$$

$\beta_1, \beta_2, \ldots, \beta_m$ are two irreducible bases of the same module $\mathfrak{a}$, the fact that both generate $\mathfrak{a}$ means that

$$\alpha_k = \sum_{\iota=1}^{m} p_k^{(\iota)} \beta_\iota, \quad \beta_k = \sum_{\iota=1}^{m} q_\iota^{(k)} \alpha_\iota,$$

where the coefficients p, q are polynomial functions of z. But this implies

$$\sum_{\iota=1}^{m} q_\iota^{(k)} p_\iota^{(h)} = 0 \quad \text{or} \quad 1$$

according as h differs from k or not, whence

$$\sum \pm p_1^{(1)} p_2^{(2)} \cdots p_m^{(m)} \cdot \sum \pm q_1^{(1)} q_2^{(2)} \cdots q_m^{(m)} = 1$$

and since both determinants are polynomial functions of z they must both be *constant*.

3. *Definition.* A module $\mathfrak{a}$ is said to be *divisible* by a module $\mathfrak{b}$, or $\mathfrak{b}$ is said to be a *divisor* of $\mathfrak{a}$, or $\mathfrak{a}$ a *multiple* of $\mathfrak{b}$ ($\mathfrak{b}$ goes into $\mathfrak{a}$), when each function in $\mathfrak{a}$ also belongs to $\mathfrak{b}$. We call $\mathfrak{b}$ a *strict*[20] *divisor* of $\mathfrak{a}$ when $\mathfrak{a}$ is divisible by $\mathfrak{b}$ but not identical to $\mathfrak{b}$.[21]

An immediate consequence of these definitions is:

If $\mathfrak{a}$ is divisible by $\mathfrak{b}$, and $\mathfrak{b}$ is divisible by $\mathfrak{c}$, then $\mathfrak{a}$ is also divisible by $\mathfrak{c}$.

4. *Definition.* The collection $\mathfrak{m}$ of all the functions belonging to both modules $\mathfrak{a}$ and $\mathfrak{b}$, unless it consists only of the "zero" function, is a module (according to the general definition) called the *least common multiple of* $\mathfrak{a}$ and $\mathfrak{b}$, because any multiple of both $\mathfrak{a}$ and $\mathfrak{b}$ is also a multiple of $\mathfrak{m}$. Similarly, the least common multiple of arbitrarily many modules $\mathfrak{a}, \mathfrak{b}, \mathfrak{c}, \ldots$ is the collection of functions that belong to all of $\mathfrak{a}, \mathfrak{b}, \mathfrak{c}, \ldots$. One arrives at the same thing by successively replacing any two of the modules $\mathfrak{a}, \mathfrak{b}, \mathfrak{c}, \ldots$ by their least common multiple.

5. *Definition.* If α is an arbitrary function in $\mathfrak{a}$ and β is an arbitrary function in $\mathfrak{b}$, then the collection of all functions of the form $\alpha + \beta$ forms a module $\mathfrak{d}$ called the *greatest common divisor of the modules $\mathfrak{a}$ and $\mathfrak{b}$*. When $\mathfrak{a}$ and $\mathfrak{b}$ are finitely generated modules, so is their greatest common divisor. Namely, if

$$\mathfrak{a} = [\alpha_1, \alpha_2, \ldots, \alpha_r], \quad \mathfrak{b} = [\beta_1, \beta_2, \ldots, \beta_s]$$

then

$$\mathfrak{d} = [\alpha_1, \alpha_2, \ldots, \alpha_r, \beta_1, \beta_2, \ldots, \beta_s].$$

It follows from the definition of divisibility that $\mathfrak{d}$ is a divisor of both $\mathfrak{a}$ and $\mathfrak{b}$. Conversely, if $\mathfrak{d}'$ is a divisor of both $\mathfrak{a}$ and $\mathfrak{b}$ then the functions α and β, hence also the functions $\alpha + \beta$, belong to $\mathfrak{d}'$. Therefore $\mathfrak{d}$ is divisible by $\mathfrak{d}'$.

The definition of the greatest common divisor of an arbitrary number of modules builds on this in the obvious way.

---

[20]I am using "strict" as a translation of the word "echt" here, rather than "proper," in order to use "proper/improper" as the translations of "eigentlich/uneigentlich," which occur from §21 onwards. (Translator's note.)

[21]The concept of divisibility of modules is intuitively opposite to that of numbers, inasmuch as the divisor contains a greater quantity of functions than the multiple. (Translator's note: However, the idea that "to divide is to contain" can be easily reconciled with ordinary concept of divisibility, if one bears in mind that b divides a if and only if the multiples of b include the multiples of a.)

6. *Definition.* If $\mathfrak{a}$ is a module, α is an arbitrary function in $\mathfrak{a}$, and μ is an arbitrary function in Ω, then the product $\mu\mathfrak{a}$ or $\mathfrak{a}\mu$ is the collection of all functions $\mu\alpha$, where α belongs to $\mathfrak{a}$. It is again a module. If

$$\mathfrak{a} = [\alpha_1, \alpha_2, \ldots, \alpha_r]$$

is a finitely generated module, then

$$\mu\mathfrak{a} = [\mu\alpha_1, \mu\alpha_2, \ldots, \mu\alpha_r]$$

is likewise a finitely generated module, and $\mu\mathfrak{a} = \mu\mathfrak{b}$ implies $\mathfrak{a} = \mathfrak{b}$ when μ is nonzero.

7. *Definition.* If $\mathfrak{a}, \mathfrak{b}$ are two modules, and α, β are any functions in $\mathfrak{a}, \mathfrak{b}$ respectively, then the product

$$\mathfrak{a}\mathfrak{b} = \mathfrak{b}\mathfrak{a} = \mathfrak{c}$$

is the collection of all products $\alpha\beta$ and all sums of such products, i.e., all the functions that can be represented by the expression

$$\gamma = \sum \alpha\beta.$$

This system of functions also constitutes a module, and indeed it is finitely generated when $\mathfrak{a}$ and $\mathfrak{b}$ are. If $\mathfrak{a}$ and $\mathfrak{b}$ are defined as in 5, then the rs functions $\alpha_\iota\beta_\kappa$, if they also form an irreducible collection, are a basis of $\mathfrak{c}$. A product of arbitrarily many modules $\mathfrak{a}, \mathfrak{b}, \mathfrak{c}, \ldots$ is built up in the obvious way, and satisfies the fundamental theorem on exchangeability of factors in multiplication. If the individual terms in such a product are all equal to $\mathfrak{a}$, and are m in number, then we denote it by $\mathfrak{a}^m$, and it has the property

$$\mathfrak{a}^{m+m'} = \mathfrak{a}^m\mathfrak{a}^{m'}.$$

In general the product $\mathfrak{a}\mathfrak{b}$ is *not* divisible by $\mathfrak{a}$. However, we have the following theorem, whose proof is immediate from the definition

If $\mathfrak{a}$ is divisible by $\mathfrak{a}_1$, and $\mathfrak{b}$ by $\mathfrak{b}_1$, then $\mathfrak{a}\mathfrak{b}$ is divisible by $\mathfrak{a}_1\mathfrak{b}_1$.

8. *Definition.* The quotient $\frac{\mathfrak{b}}{\mathfrak{a}}$ of two modules $\mathfrak{a}, \mathfrak{b}$ is the collection of all those functions γ with the property that $\gamma\mathfrak{a}$ is divisible by $\mathfrak{b}$. This quotient, unless it consists of the single function "zero," is itself a module $\mathfrak{c}$, as is immediately clear from the definition. The product $\frac{\mathfrak{b}}{\mathfrak{a}} \cdot \mathfrak{a}$ is always divisible by $\mathfrak{b}$, though not always equal to $\mathfrak{b}$.

§5. Congruences

SUMMARY AND COMMENTS

The reason for the centrality of the module concept (and the reason for its name) is that a module $\mathfrak{a}$ admits the relation of *congruence modulo* $\mathfrak{a}$. If we say that $\alpha \equiv \beta \pmod{\mathfrak{a}}$ if $\alpha - \beta \in \mathfrak{a}$, then the basic properties of congruence follow from the closure properties of modules.

In algebraic number theory the number of congruence classes is typically finite, but in algebraic function theory the congruence classes form a *vector space*[22] over $\mathbb{C}$. That is, there are algebraic functions $\lambda_1, \lambda_2, \ldots, \lambda_m$ such that the functions

$$c_1\lambda_1 + c_2\lambda_2 + \cdots + c_m\lambda_m$$

represent the distinct congruence classes (mod $\mathfrak{a}$) as $c_1, c_2, \ldots, c_m$ run through the distinct m-tuples of complex numbers. (Dedekind and Weber say that $\lambda_1, \lambda_2, \ldots, \lambda_m$ form a "complete system of remainders.")

The same is true when a module $\mathfrak{b}$ is considered *relative* to a module $\mathfrak{a}$—the congruence classes of $\mathfrak{b}$ modulo $\mathfrak{a}$ form a vector space. If the functions $\lambda_1, \lambda_2, \ldots, \lambda_m$ are linearly independent over $\mathbb{C}$ then they form a *basis* of this system, which is then written as $(\lambda_1, \lambda_2, \ldots, \lambda_m)$.

Two functions α, β are called *congruent* modulo the module $\mathfrak{a}$,

$$\alpha \equiv \beta \pmod{\mathfrak{a}},$$

when the difference $\alpha - \beta$ belongs to the module $\mathfrak{a}$.

The following theorems are immediate from the definition:

1. If $\alpha \equiv \beta, \beta \equiv \gamma \pmod{\mathfrak{a}}$ then $\alpha \equiv \gamma \pmod{\mathfrak{a}}$.

2. If $\mathfrak{d}$ is any divisor of $\mathfrak{a}$ then $\alpha \equiv \beta \pmod{\mathfrak{a}}$ implies $\alpha \equiv \beta \pmod{\mathfrak{d}}$.

3. If $\alpha \equiv \beta \pmod{\mathfrak{a}}$ and μ is any function in Ω, then $\mu\alpha \equiv \mu\beta \pmod{\mu\mathfrak{a}}$, and conversely, the latter congruence follows from the former when μ is nonzero.

4. If $\alpha \equiv \beta, \alpha_1 \equiv \beta_1 \pmod{\mathfrak{a}}$ then $\alpha \pm \alpha_1 \equiv \beta \pm \beta_1 \pmod{\mathfrak{a}}$.

If $\lambda_1, \lambda_2, \ldots, \lambda_n$ are given functions in Ω, and $c_1, c_2, \ldots, c_m$ are *arbitrary constants*, then the collection of all functions of the form

$$c_1\lambda_1 + c_2\lambda_2 + \cdots + c_m\lambda_m$$

is called a *vector space* and is denoted by $(\lambda_1, \lambda_2, \ldots, \lambda_m)$. The system of functions $\lambda_1, \lambda_2, \ldots, \lambda_m$ is called the *basis of the vector space*. The functions $\lambda_1, \lambda_2, \ldots, \lambda_m$ are said to be *linearly independent* or a *linearly irreducible* system, if an equation (identity) of the form

$$c_1\lambda_1 + c_2\lambda_2 + \cdots + c_m\lambda_m = 0$$

holds only when the constant coefficients $c_1, c_2, \ldots, c_m$ all vanish.

We have the theorem that *each vector space has a linearly irreducible basis.* Because if $c_1\lambda_1 + c_2\lambda_2 + \cdots + c_m\lambda_m = 0$ and c_1 is nonzero, then the vector space $(\lambda_1, \lambda_2, \ldots, \lambda_m)$ is identical with the vector space $(\lambda_2, \lambda_3, \ldots, \lambda_m)$, whose basis contains one function less. If the latter is not linearly irreducible, then one can continue similarly. Here too we shall understand a "basis" to be an irreducible basis from now on. The number of functions in an irreducible basis of a vector space is always the same and is called the *dimension* of the vector space. If the dimension is m, the vector space is said to be *m-tuple.* Any m functions in an m-tuple vector space form an irreducible basis if and only if they are linearly independent.

---

[22]Dedekind and Weber call it a *Schaar*. I have opted for the anachronistic, but accurate, translation "vector space" because any 19th-century English equivalent is likely to be misleading. (Translator's note.)

The functions $\lambda_1, \lambda_2, \ldots, \lambda_m$ are called *linearly independent modulo the module* $\mathfrak{a}$ if a congruence of the form

$$c_1\lambda_1 + c_2\lambda_2 + \cdots + c_m\lambda_m \equiv 0 \pmod{\mathfrak{a}}$$

holds only when the constant coefficients $c_1, c_2, \ldots, c_m$ vanish. Then two sums of the form $\sum c_\iota \lambda_\iota$ with different values of the constant coefficients c_ι are always incongruent modulo the module $\mathfrak{a}$.

Now let $\mathfrak{a}$ and $\mathfrak{b}$ be two modules, and suppose *first* that in $\mathfrak{b}$ only finitely many functions $\lambda_1, \lambda_2, \ldots, \lambda_m$ are linearly independent modulo $\mathfrak{a}$. Then each function β in $\mathfrak{b}$ satisfies exactly one congruence of the form

$$\beta \equiv c_1\lambda_1 + c_2\lambda_2 + \cdots + c_m\lambda_m \pmod{\mathfrak{a}}$$

with constant coefficients $c_1, c_2, \ldots, c_m$. The vector space $(\lambda_1, \lambda_2, \ldots, \lambda_m)$ can thus be called a *complete system of remainders*[23] of $\mathfrak{b}$ *modulo* $\mathfrak{a}$, *and* $\lambda_1, \lambda_2, \ldots, \lambda_m$ its *basis*. One can write this symbolically as

$$\mathfrak{b} \equiv (\lambda_1, \lambda_2, \ldots, \lambda_m) \pmod{\mathfrak{a}}.$$

If one chooses any system of m functions $\lambda_1', \lambda_2', \ldots, \lambda_m'$ in $\mathfrak{b}$, then there are m congruences

$$\lambda_k' \equiv \sum_\iota k_{h,\iota}\lambda_\iota \pmod{\mathfrak{a}}$$

with constants $k_{h,\iota}$, and this system is a basis of a complete remainder system of $\mathfrak{b}$ modulo $\mathfrak{a}$ if and only if the determinant

$$\sum \pm k_{1,1}k_{2,2}\cdots k_{m,m}$$

is nonzero.

§6. The norm of one module relative to another

Summary and Comments

In algebraic number theory, the norm of a module $\mathfrak{b}$ relative to a module $\mathfrak{a}$, $(\mathfrak{b}, \mathfrak{a})$, is simply the number of congruence classes of $\mathfrak{b}$ modulo $\mathfrak{a}$.

In algebraic function theory, as we saw in the previous section, the congruence classes of $\mathfrak{b}$ modulo $\mathfrak{a}$ form a vector space $(\lambda_1, \lambda_2, \ldots, \lambda_m)$. To define a relative norm $(\mathfrak{b}, \mathfrak{a})$, Dedekind and Weber associate a determinant with $(\lambda_1, \lambda_2, \ldots, \lambda_m)$—the characteristic polynomial of multiplication by z on $\mathfrak{b}/\mathfrak{a}$—which is of degree m.

The multiplicative property of determinants then yields a multiplicative property of the relative norm: if $\mathfrak{c}$ divides $\mathfrak{b}$ and $\mathfrak{b}$ divides $\mathfrak{a}$ then $(\mathfrak{c}, \mathfrak{a}) = (\mathfrak{c}, \mathfrak{b})(\mathfrak{b}, \mathfrak{a})$.

---

[23]I have translated the German word "Rest" as "remainder" here, rather than the more common "residue," to avoid conflict with the word "residue" that Dedekind and Weber use later, inspired by the concept of residue in complex analysis. (Translator's note.)

If $(\lambda_1, \lambda_2, \ldots, \lambda_m)$ is a complete remainder system of $\mathfrak{b}$ modulo $\mathfrak{a}$ then, since $z\mathfrak{b}$ is divisible by $\mathfrak{b}$, there is a system of m^2 constants $c_{h,k}$ satisfying the congruences

$$\left. \begin{aligned} z\lambda_1 &\equiv c_{1,1}\lambda_1 + c_{2,1}\lambda_2 + \cdots + c_{m,1}\lambda_m \\ z\lambda_2 &\equiv c_{1,2}\lambda_1 + c_{2,2}\lambda_2 + \cdots + c_{m,2}\lambda_m \\ &\cdots\cdots\cdots\cdots\cdots\cdots\cdots\cdots\cdots\cdots\cdots \\ z\lambda_m &\equiv c_{1,m}\lambda_1 + c_{2,m}\lambda_2 + \cdots + c_{m,m}\lambda_m \end{aligned} \right\} \pmod{\mathfrak{a}}$$

and by solving this system one sees that each function λ, and hence each function β in the module $\mathfrak{b}$, may be converted into one in the module $\mathfrak{a}$ by multiplying by the mth degree polynomial in z

$$(\mathfrak{b}, \mathfrak{a}) = (-1)^m \begin{vmatrix} c_{1,1} - z & c_{2,1} & \cdots & c_{m,1} \\ c_{1,2} & c_{2,2} - z & \cdots & c_{m,2} \\ \cdots & \cdots & \cdots & \cdots \\ c_{1,m} & c_{2,m} & \cdots & c_{m,m} - z \end{vmatrix}.$$

It follows easily from the multiplication theorem for determinants that this function $(\mathfrak{b}, \mathfrak{a})$ is independent of the choice of basis $\lambda_1, \lambda_2, \ldots, \lambda_m$, and hence dependent only on the two modules $\mathfrak{a}, \mathfrak{b}$. We call it the *norm of* $\mathfrak{a}$ *relative to* $\mathfrak{b}$.

If each function in $\mathfrak{b}$ also belongs to $\mathfrak{a}$, so that $\mathfrak{b}$ is divisible by $\mathfrak{a}$, then we set $m = 0$ and $(\mathfrak{b}, \mathfrak{a}) = 1$. If, contrary to the assumption above, there are more than a finite number of functions in $\mathfrak{b}$ that are linearly independent modulo $\mathfrak{a}$, then it is agreed that $(\mathfrak{b}, \mathfrak{a}) = 0$.

1. If $\mathfrak{m}$ is the least common multiple, and $\mathfrak{d}$ is the greatest common divisor, of $\mathfrak{a}$ and $\mathfrak{b}$, then each congruence between functions in $\mathfrak{b}$ modulo $\mathfrak{a}$ is equivalent to the same congruence modulo $\mathfrak{m}$. On the other hand, each function in $\mathfrak{b}$ is congruent to a function in $\mathfrak{d}$ modulo $\mathfrak{a}$, and conversely each function in $\mathfrak{d}$ is congruent to a function in $\mathfrak{b}$ modulo $\mathfrak{a}$. These remarks immediately yield the important theorem

$$(\mathfrak{b}, \mathfrak{a}) = (\mathfrak{b}, \mathfrak{m}) = (\mathfrak{d}, \mathfrak{a}),$$

which also remains true when $(\mathfrak{b}, \mathfrak{a}) = 0$.

2. *If the module* $\mathfrak{a}$ *is divisible by the module* $\mathfrak{b}$*, and* $\mathfrak{b}$ *is divisible by the third module* $\mathfrak{c}$*, then*

$$(\mathfrak{c}, \mathfrak{a}) = (\mathfrak{c}, \mathfrak{b})(\mathfrak{b}, \mathfrak{a}).$$

The theorem is obviously correct when one of the two norms $(\mathfrak{c}, \mathfrak{b}), (\mathfrak{b}, \mathfrak{a})$ is zero. If this is not the case and if

$$\mathfrak{c} \equiv (\rho_1, \rho_2, \ldots, \rho_r) \pmod{\mathfrak{b}},$$
$$\mathfrak{b} \equiv (\lambda_1, \lambda_2, \ldots, \lambda_s) \pmod{\mathfrak{a}},$$

then the functions $\rho_1, \rho_2, \ldots, \rho_r, \lambda_1, \lambda_2, \ldots, \lambda_s$ together are linearly independent modulo $\mathfrak{a}$, because if

$$\sum_\iota c_\iota \rho_\iota + \sum_\iota c'_\iota \lambda_\iota \equiv 0 \pmod{\mathfrak{a}},$$

then it follows, since $\mathfrak{a}$ is divisible by $\mathfrak{b}$ and the functions λ_ι are in $\mathfrak{b}$, that

$$\sum_\iota c_\iota \rho_\iota \equiv 0 \pmod{\mathfrak{b}}, \text{ hence } c_\iota = 0,$$

$$\sum_\iota c'_\iota \lambda_\iota \equiv 0 \pmod{\mathfrak{a}}, \text{ hence } c'_\iota = 0.$$

Also, since each function γ in $\mathfrak{c}$ satisfies a congruence of the form

$$\gamma \equiv \sum c_\iota \rho_\iota + \sum c'_\iota \lambda_\iota \pmod{\mathfrak{a}},$$

the $(r+s)$-tuple vector space $(\rho_1, \rho_2, \ldots, \rho_r, \lambda_1, \lambda_2, \ldots, \lambda_s)$ is a complete remainder system of $\mathfrak{c}$ modulo $\mathfrak{a}$, or

$$\mathfrak{c} \equiv (\rho_1, \rho_2, \ldots, \rho_r, \lambda_1, \lambda_2, \ldots, \lambda_s) \pmod{\mathfrak{a}}.$$

Therefore if

$$z\rho_1 = e_{1,1}\rho_1 + \cdots + e_{r,1}\rho_r + \beta_1,$$

$$\cdots\cdots\cdots\cdots\cdots\cdots\cdots\cdots\cdots\cdots\cdots$$

$$z\rho_r = e_{1,r}\rho_1 + \cdots + e_{r,r}\rho_r + \beta_r,$$

where the $e_{\iota,\kappa}$ are constants and the β_i are functions in $\mathfrak{b}$, so that

$$(\mathfrak{c}, \mathfrak{b}) = (-1)^r \begin{vmatrix} e_{1,1} - z & \cdots & e_{r,1} \\ \cdots & \cdots & \cdots \\ e_{1,r} & \cdots & e_{r,r} - z, \end{vmatrix},$$

and if also

$$\left. \begin{array}{l} \beta_1 \equiv h_{1,1}\lambda_1 + \cdots + h_{s,1}\lambda_s \\ \cdots\cdots\cdots\cdots\cdots\cdots\cdots \\ \beta_r \equiv h_{1,r}\lambda_1 + \cdots + h_{s,r}\lambda_s \\ z\lambda_1 \equiv c_{1,1}\lambda_1 + \cdots + c_{s,1}\lambda_s \\ \cdots\cdots\cdots\cdots\cdots\cdots\cdots \\ z\lambda_s \equiv c_{1,s}\lambda_1 + \cdots + c_{s,s}\lambda_s \end{array} \right\} \pmod{\mathfrak{a}}$$

with constant coefficients $h_{\iota,\kappa}, c_{\iota,\kappa}$, so that

$$(\mathfrak{b}, \mathfrak{a}) = (-1)^s \begin{vmatrix} c_{1,1} - z & \cdots & c_{s,1} \\ \cdots & \cdots & \cdots \\ c_{1,s} & \cdots & c_{s,s} - z \end{vmatrix}$$

then it follows that

$$\left. \begin{array}{l} z\rho_1 \equiv e_{1,1}\rho_1 + \cdots + e_{r,1}\rho_r + h_{1,1}\lambda_1 + \cdots + h_{s,1}\lambda_s \\ \cdots\cdots\cdots\cdots\cdots\cdots\cdots\cdots\cdots\cdots\cdots\cdots\cdots \\ z\rho_r \equiv e_{1,r}\rho_1 + \cdots + e_{r,r}\rho_r + h_{1,r}\lambda_1 + \cdots + h_{s,r}\lambda_s \\ z\lambda_1 \equiv \qquad\qquad\qquad\qquad\quad c_{1,1}\lambda_1 + \cdots + c_{s,1}\lambda_s \\ \cdots\cdots\cdots\cdots\cdots\cdots\cdots\cdots\cdots\cdots\cdots\cdots\cdots \\ z\lambda_s \equiv \qquad\qquad\qquad\qquad\quad c_{1,s}\lambda_1 + \cdots + c_{s,s}\lambda_s \end{array} \right\} \pmod{\mathfrak{a}}$$

whence

$$(\mathfrak{c}, \mathfrak{a}) = (-1)^{r+s} \begin{vmatrix} e_{1,1} - z & \cdots & e_{r,1} & h_{1,1} & \cdots & h_{s,1} \\ \cdots & \cdots & \cdots & \cdots & & \cdots \\ e_{1,r} & \cdots & e_{r,r} - z & h_{1,r} & \cdots & h_{s,r} \\ 0 & \cdots & 0 & c_{1,1} - z & \cdots & c_{s,1} \\ \cdots & \cdots & \cdots & \cdots & \cdots & \cdots \\ 0 & \cdots & 0 & c_{1,s} & \cdots & c_{s,s} - z \end{vmatrix} = (\mathfrak{c}, \mathfrak{b})(\mathfrak{b}, \mathfrak{a}).$$

3. When the basis functions $\beta_1, \beta_2, \ldots, \beta_s$ of a finitely generated module $\mathfrak{b} = [\beta_1, \beta_2, \ldots, \beta_s]$ can be converted to functions in a module $\mathfrak{a}$ by multiplication by nonzero polynomials in z, the norm $(\mathfrak{b}, \mathfrak{a})$ of $\mathfrak{a}$ relative to $\mathfrak{b}$ is nonzero. At the same

time, the least common multiple of $\mathfrak{a}$ and $\mathfrak{b}$ is a finitely generated module $\mathfrak{m}$, an irreducible basis of which is

$$\mu_1 = a_{1,1}\beta_1,$$
$$\mu_2 = a_{1,2}\beta_1 + a_{2,2}\beta_2,$$
$$\dots\dots\dots\dots\dots\dots\dots$$
$$\mu_s = a_{1,s}\beta_1 + a_{2,s}\beta_2 + \cdots + a_{s,s}\beta_s,$$

where the coefficients $a_{\iota,\kappa}$ are polynomial functions of z, and indeed such that

$$(\mathfrak{b}, \mathfrak{a}) = a_{1,1}a_{2,2}\cdots a_{s,s}.$$

To prove this important theorem we let $\mathfrak{a}_1$ be the greatest common divisor of $\mathfrak{a}$ and $[\beta_1]$, let $\mathfrak{a}_2$ be the greatest common divisor of $\mathfrak{a}_1$ and $[\beta_2]$, etc., so that $\mathfrak{a}_r$ is the collection of all functions of the form

$$\alpha_r = \alpha + y_1\beta_1 + \cdots + y_r\beta_r,$$

where α is a function in $\mathfrak{a}$ and $y_1, \ldots, y_r$ are polynomial functions of z. Then $\mathfrak{a}_s$ is the greatest common divisor of $\mathfrak{a}$ and $\mathfrak{b}$. Now since each module $\mathfrak{a}_r$ is divisible by the next, $\mathfrak{a}_{r+1}$, it follows from 1 and 2 that

$$(\mathfrak{b}, \mathfrak{a}) = (\mathfrak{a}_s, \mathfrak{a}) = (\mathfrak{a}_s, \mathfrak{a}_{s-1})(\mathfrak{a}_{s-1}, \mathfrak{a}_{s-2})\cdots(\mathfrak{a}_1, \mathfrak{a}),$$

and hence it is a matter of determining $(\mathfrak{a}_r, \mathfrak{a}_{r-1})$. But

$$\alpha_r \equiv \alpha_{r-1} + y_r\beta_r \equiv y_r\beta_r \pmod{\mathfrak{a}_{r-1}},$$

and by hypothesis there is a nonzero polynomial function x_r of z for which

$$x_r\beta_r \equiv 0 \pmod{\mathfrak{a}}$$

and hence also

$$x_r\beta_r \equiv 0 \pmod{\mathfrak{a}_{r-1}}.$$

Now if $a_{r,r}$ is a function of minimal degree m_r among all the functions x_r satisfying the latter congruence, and also such that the coefficient of the highest power of z is 1, then all other functions x_r satisfying this congruence are divisible by $a_{r,r}$. This is because

$$(x_r - qa_{r,r})\beta_r \equiv 0 \pmod{\mathfrak{a}_{r-1}}$$

for any polynomial q, hence if x_r is not divisible by $a_{r,r}$, q may be chosen so that $x_r - qa_{r,r}$ is of lower degree than $a_{r,r}$, contrary to hypothesis.

Thus, if one sets

$$y_r = qa_{r,r} + b_{r,r}$$

and determines q so that the degree of $b_{r,r}$ is smaller than m_r, it follows that

$$\alpha_r \equiv b_{r,r}\beta_r \pmod{\mathfrak{a}_{r-1}},$$

whence

$$\mathfrak{a}_r \equiv (\beta_r, z\beta_r, \ldots, z^{m_r-1}\beta_r) \pmod{\mathfrak{a}_{r-1}}.$$

Consequently, when one for the moment sets

$$a_{r,r} = c_0 + c_1 z + \cdots + c_{m_r-1}z^{m_r-1} + z^{m_r},$$
$$\lambda_k = z^{k-1}\beta_r,$$

it follows that

$$z\lambda_1 = \lambda_2, \quad z\lambda_2 = \lambda_3, \quad \ldots$$

$$z\lambda_{m_r} \equiv -c_0\lambda_1 - c_1\lambda_2 - \cdots - c_{m_r-1}\lambda_{m_r} \pmod{\mathfrak{a}_{r-1}},$$

hence

$$(\mathfrak{a}_r, \mathfrak{a}_{r-1}) = (-1)^{m_r}
\begin{vmatrix}
-z & 1 & 0 & \ldots & & 0 \\
0 & -z & 1 & \ldots & & 0 \\
\ldots & \ldots & \ldots & \ldots & & \ldots \\
0 & 0 & 0 & & & 1 \\
-c_0 & -c_1 & -c_2 & & -c_{m_r-1} & -z
\end{vmatrix} = a_{r,r}.$$

The result, as in 2, is that the function system

$$\beta_1, \quad z\beta_1, \quad \ldots, \quad z^{m_1-1}\beta_1,$$

$$\beta_2, \quad z\beta_2, \quad \ldots, \quad z^{m_2-1}\beta_2,$$

$$\cdots\cdots\cdots\cdots\cdots\cdots\cdots\cdots\cdots\cdots$$

$$\beta_s, \quad z\beta_s, \quad \ldots, \quad z^{m_s-1}\beta_s,$$

is a basis for a complete remainder system of $\mathfrak{b}$ modulo $\mathfrak{a}$, and that

$$(\mathfrak{b}, \mathfrak{a}) = a_{1,1}a_{2,2}\cdots a_{s,s}$$

is of degree $m = m_1 + m_2 + \cdots + m_s$.

Now, since $a_{r,r}\beta_r \equiv 0 \pmod{\mathfrak{a}_{r-1}}$, a function μ_r in $\mathfrak{a}$ and polynomial functions $a_{k,r}$ may be determined so that

$$\mu_r = a_{1,r}\beta_1 + a_{2,r}\beta_2 + \cdots + a_{r,r}\beta_r.$$

The functions

$$\mu_1 = a_{1,1}\beta_1,$$

$$\mu_2 = a_{1,2}\beta_1 + a_{2,2}\beta_2,$$

$$\cdots\cdots\cdots\cdots\cdots\cdots\cdots\cdots\cdots$$

$$\mu_s = a_{1,s}\beta_1 + a_{2,s}\beta_2 + \cdots + a_{s,s}\beta_s$$

determined in this way are, since none of the functions $a_{1,1}, \ldots, a_{s,s}$ vanishes, rationally independent. At the same time, they are all in both $\mathfrak{a}$ and $\mathfrak{b}$, hence also in the least common multiple $\mathfrak{m}$ of these two modules. It remains to prove that they form a basis of $\mathfrak{m}$.

Let $\mathfrak{m}_r$ be the least common multiple of $\mathfrak{a}$ and $[\beta_1, \beta_2, \ldots, \beta_r]$, $\mathfrak{m}_s = \mathfrak{m}$, so that each of the modules $\mathfrak{m}_1, \mathfrak{m}_2, \ldots, \mathfrak{m}_s$ is divisible by the next, and hence also by $\mathfrak{m}$, and let

$$\nu_r = z_1\beta_1 + z_2\beta_2 + \cdots + z_r\beta_r$$

be a function in $\mathfrak{m}_r$, hence also in $\mathfrak{a}$.

This means

$$z_r\beta_r \equiv 0 \pmod{\mathfrak{a}_{r-1}},$$

hence

$$z_r = x_r a_{r,r},$$

where x_r is a polynomial function. Therefore

$$\nu_r - x_r\mu_r \equiv 0 \pmod{\mathfrak{m}_{r-1}}, \quad \nu_1 - x_1\mu_1 = 0,$$

whence

$$\nu_r = x_1\mu_1 + x_2\mu_2 + \cdots + x_r\mu_r,$$

and so

$$\mathfrak{m}_r = [\mu_1, \mu_2, \ldots, \mu_r],$$
$$\mathfrak{m} = [\mu_1, \mu_2, \ldots, \mu_s].$$

Q.E.D.

Thus an irreducible basis of the module $\mathfrak{m}$ contains precisely as many functions as an irreducible basis of $\mathfrak{b}$. If one chooses another basis $\mu_1', \mu_2', \ldots, \mu_s'$ in place of $\mu_1, \mu_2, \ldots, \mu_s$, then $\mu_1', \mu_2', \ldots, \mu_s'$ may be expressed in the form

$$\mu_k' = a_{1,k}'\beta_1 + a_{2,k}'\beta_2 + \cdots + a_{s,k}'\beta_s$$

with polynomial coefficients $a_{\iota,\kappa}'$, and it follows from §4, 2 that

$$(\mathfrak{b}, \mathfrak{a}) = \text{const.} \sum \pm a_{1,1}' a_{2,2}' \cdots a_{s,s}'.$$

4. In particular, if we make the assumption that $\mathfrak{a}$ is likewise a finitely generated module, possessing an irreducible basis with the same number of functions as the basis of $\mathfrak{b}$, and if $\mathfrak{a}$ is divisible by $\mathfrak{b}$ as well, and

$$\mathfrak{a} = [\alpha_1, \alpha_2, \ldots, \alpha_s]$$

then the polynomial functions $b_{\iota,k}$ may be determined in such a way that

$$\alpha_k = b_{1,k}\beta_1 + b_{2,k}\beta_2 + \cdots + b_{s,k}\beta_s.$$

Then the hypothesis of 3, that the functions β_ι are convertible to functions in $\mathfrak{a}$ by multiplication by polynomial functions of z, is satisfied, as one sees by solution of this equation system. At the same time, $\mathfrak{a}$ is itself the least common multiple of $\mathfrak{a}$ and $\mathfrak{b}$, and therefore

$$(\mathfrak{b}, \mathfrak{a}) = \text{const.} \sum \pm b_{1,1}b_{2,2} \cdots b_{n,n}.$$

5. If $\mathfrak{m}$ is the least common multiple of two modules $\mathfrak{a}$, $\mathfrak{b}$ and if ν is an arbitrary function in Ω then, as follows easily from the definition, $\nu\mathfrak{m}$ is the least common multiple of $\nu\mathfrak{a}$ and $\nu\mathfrak{b}$. If $(\mathfrak{b}, \mathfrak{a}) = 0$ then $(\nu\mathfrak{b}, \nu\mathfrak{a}) = 0$ also. But if $(\mathfrak{b}, \mathfrak{a})$ and ν are nonzero, one gets

$$(\nu\mathfrak{b}, \nu\mathfrak{a}) = (\mathfrak{b}, \mathfrak{a})$$

when the basis functions μ_ι, β_ι of $\mathfrak{m}$ and $\mathfrak{b}$ in 3 are replaced by $\nu\mu_\iota, \nu\beta_\iota$.

§7. The ideals in $\mathfrak{o}$

SUMMARY AND COMMENTS

An ideal in the ring $\mathfrak{o}$ of integral algebraic functions is defined, as usual, as a set closed under sum and difference, and under products with arbitrary members of $\mathfrak{o}$. As with modules in general, "to divide is to contain," so $\mathfrak{o}$ itself divides every ideal. Also as usual, a *principal ideal* is one consisting of all the multiples of a fixed member of $\mathfrak{o}$.

The *norm* $N(\mathfrak{a})$ of an ideal $\mathfrak{a}$ is the relative norm $(\mathfrak{o}, \mathfrak{a})$. As shown in the previous section, $N(\mathfrak{a})$ is a polynomial, and the degree of this polynomial is called the *degree of* $\mathfrak{a}$. $N(\mathfrak{a}) = 1$ if and only if $\mathfrak{a} = \mathfrak{o}$, and if $\alpha \in \mathfrak{a}$ then $N(\mathfrak{a})$ divides $N(\alpha)$.

Congruences modulo an ideal $\mathfrak{a}$ have the following convenient property, which does *not* hold for modules in general: $\mu \equiv \mu_1$, $\nu \equiv \nu_1$ (mod $\mathfrak{a}$) implies $\mu\nu \equiv \mu_1\nu_1$ (mod $\mathfrak{a}$).

A system $\mathfrak{a}$ of *integral algebraic* functions of z in the field Ω is called an *ideal* when it satisfies the following conditions:

  I. The sum and difference of any two functions in $\mathfrak{a}$ are again functions in $\mathfrak{a}$.
  II. The product of any function in $\mathfrak{a}$ with any function in $\mathfrak{o}$ (§3) is again a function in $\mathfrak{a}$.

Every ideal is at the same time a module, and all terms and notations defined for modules can be applied to ideals.

The module $\mathfrak{o}$ (the system of all integral algebraic functions of z) is itself an ideal, and *each ideal is divisible by* $\mathfrak{o}$. Likewise, when μ is any nonzero function in $\mathfrak{o}$, the module $\mathfrak{o}\mu$ (the system of all integral algebraic functions divisible by μ) is an ideal. Such an ideal is called a *principal ideal*. If $\omega_1, \omega_2, \ldots, \omega_n$ is a basis of $\mathfrak{o}$ then

$$\mathfrak{o}\mu = [\omega_1\mu, \omega_2\mu, \ldots, \omega_n\mu]$$

and $\mathfrak{o}\mu$ is the least common multiple of $\mathfrak{o}$ and $\mathfrak{o}\mu$. It then follows from §6, 4 and definition (4) of §2 that

(1) $$(\mathfrak{o}, \mathfrak{o}\mu) = \text{const.} N(\mu),$$

and consequently it is nonzero.

If $\mathfrak{a}$ is any ideal and α is an arbitrary function in $\mathfrak{a}$ then (by II) the principal ideal $\mathfrak{o}\alpha$ is divisible by $\mathfrak{a}$. It follows by §6, 2 that

(2) $$(\mathfrak{o}, \mathfrak{o}\alpha) = (\mathfrak{o}, \mathfrak{a})(\mathfrak{a}, \mathfrak{o}\alpha)$$

and hence $(\mathfrak{o}, \mathfrak{a})$ is also nonzero. Now since $\mathfrak{a}$ is again the least common multiple of $\mathfrak{a}$ and $\mathfrak{o}$, it follows by §6, 3 that $\mathfrak{a}$ has an irreducible basis consisting of n integral algebraic functions $\alpha_1, \alpha_2, \ldots, \alpha_n$, which therefore also form a basis of the field Ω.

The norm of $\mathfrak{a}$ relative to $\mathfrak{o}$, i.e., the polynomial function $(\mathfrak{o}, \mathfrak{a})$ of z, will be called the *norm of the ideal* $\mathfrak{a}$ and denoted by $N(\mathfrak{a})$. The degree of this polynomial is called the *degree of the ideal* $\mathfrak{a}$.

If

$$\mathfrak{a} = [\alpha_1, \alpha_2, \ldots, \alpha_n], \quad \mathfrak{o} = [\omega_1, \omega_2, \ldots, \omega_n]$$

and

$$\alpha_1 = a_{1,1}\omega_1 + a_{2,1}\omega_2 + \cdots + a_{n,1}\omega_n,$$
$$\alpha_2 = a_{1,2}\omega_1 + a_{2,2}\omega_2 + \cdots + a_{n,2}\omega_n,$$
$$\cdots\cdots\cdots\cdots\cdots\cdots\cdots\cdots\cdots\cdots\cdots$$
$$\alpha_n = a_{1,n}\omega_1 + a_{2,n}\omega_2 + \cdots + a_{n,n}\omega_n$$

with polynomial coefficients $a_{\iota,\kappa}$, then it follows from §6, 4 that

(3) $$N(\mathfrak{a}) = \text{const.} \sum \pm a_{1,1} a_{2,2} \cdots a_{n,n}.$$

Since each function in $\mathfrak{o}$, and hence also the function "1," is converted into a function in the ideal $\mathfrak{a}$ by multiplication by $N(\mathfrak{a})$, $N(\mathfrak{a})$ is always a function in $\mathfrak{a}$.

The norm of the ideal $\mathfrak{o}$ is equal to 1, and conversely $\mathfrak{o}$ is the only ideal with this property. Also, $\mathfrak{o}$ is the only ideal that contains the function "1" (or any constant).

If α is a function in $\mathfrak{a}$ then it follows from (1), (2), (3) that

$$(4) \qquad\qquad N(\alpha) = \text{const.} N(\mathfrak{a})(\mathfrak{a}, \mathfrak{o}\alpha),$$

i.e., the norm of each function in $\mathfrak{a}$ is divisible by the norm of $\mathfrak{a}$.

Congruences modulo an ideal satisfy the following theorem, which distinguishes ideals from general modules.

If μ, μ_1, ν, ν_1 are functions in $\mathfrak{o}$ that satisfy the congruences

$$\mu \equiv \mu_1, \;\; \nu \equiv \nu_1 \pmod{\mathfrak{a}},$$

then also

$$\mu\nu \equiv \mu_1\nu_1 \pmod{\mathfrak{a}}.$$

§8. Multiplication and division of ideals

SUMMARY AND COMMENTS

Multiplication and division of ideals follow from multiplication and division of modules in §4, but with extra features due to the existence of principal ideals, and with a specific focus on *prime* ideals. As in algebraic number theory the purpose of prime ideals is to recapture the *unique prime factorization* property that holds for the "ordinary" integral elements (ordinary integers in the case of algebraic number theory and polynomials in the case of algebraic function theory).

Also involved in this theory are the least common multiple, $\mathfrak{m}$, and the greatest common divisor, $\mathfrak{d}$, of ideals $\mathfrak{a}$ and $\mathfrak{b}$. Their norms are shown to satisfy

$$N(\mathfrak{a})N(\mathfrak{b}) = N(\mathfrak{m})N(\mathfrak{d}).$$

It does not yet follow that the norm is multiplicative, but only that

$$N(\mathfrak{a})N(\mathfrak{b}) = N(\mathfrak{a})N(\mathfrak{b})$$

for relatively prime $\mathfrak{a}$ and $\mathfrak{b}$.

However, it is shown that if $\mathfrak{b}$ divides $\mathfrak{a}$ then $N(\mathfrak{b})$ divides $N(\mathfrak{a})$. It follows, by splitting an ideal into factors of minimal (degree) norm, that a prime factorization *exists* for each ideal.

The main step towards proving *uniqueness* of this prime factorization is also taken in this section. It is the *prime divisor property*: if a prime ideal $\mathfrak{p}$ divides $\mathfrak{a}\mathfrak{b}$, then $\mathfrak{p}$ divides $\mathfrak{a}$ or $\mathfrak{p}$ divides $\mathfrak{b}$.

The basic properties I and II of ideals, together with the concepts of §4, lead first of all to:

1. The least common multiple, greatest common divisor, and the product of two (or arbitrarily many) ideals are themselves ideals. Likewise, when ν is a function in $\mathfrak{o}$, and $\mathfrak{a}$ is an ideal, the product $\mathfrak{a}\nu$ is an ideal.

2. The product of ideals is divisible by each of its factors, and

$$\mathfrak{a}\mathfrak{o} = \mathfrak{a}$$

for each ideal $\mathfrak{a}$. This is because I, II imply that each function in $\mathfrak{a}\mathfrak{o}$ is also a function in $\mathfrak{a}$ and, since $\mathfrak{o}$ contains the function "1," it is true conversely that each function in $\mathfrak{a}$ is also a function in $\mathfrak{a}\mathfrak{o}$.

3. A principal ideal $\mathfrak{o}\mu$ is divisible by a principal ideal $\mathfrak{o}\nu$ if and only if the integral algebraic function μ is divisible by the integral algebraic function ν.

We now add the following definitions:

4. *Definition.* A function α in $\mathfrak{o}$ is said to be *divisible* by the ideal $\mathfrak{a}$ when the principal ideal $\mathfrak{o}\alpha$ is divisible by $\mathfrak{a}$, i.e., when α is a function in $\mathfrak{a}$.

5. *Definition.* Two ideals $\mathfrak{a}, \mathfrak{b}$ are called *relatively prime* when their greatest common divisor is $\mathfrak{o}$. The necessary and sufficient condition for this is that there exist a function α in $\mathfrak{a}$ and a function β in $\mathfrak{b}$ such that

$$\alpha + \beta = 1,$$

or in other words that there exists a function α in $\mathfrak{a}$ satisfying the congruence $\alpha \equiv 1 \pmod{\mathfrak{b}}$, or a function β in $\mathfrak{b}$ satisfying the congruence $\beta \equiv 1 \pmod{\mathfrak{a}}$.

6. *Definition.* An ideal $\mathfrak{p}$ different from $\mathfrak{o}$ is called a *prime ideal* when no ideal apart from $\mathfrak{p}$ and $\mathfrak{o}$ divides $\mathfrak{p}$.

On the basis of these definitions, we now obtain the following theorems on divisibility of ideals.

7. If $\mathfrak{a}, \mathfrak{b}$ are two ideals with least common multiple $\mathfrak{m}$ and greatest common divisor $\mathfrak{d}$, then it follows from §6, 1, 2 that

$$N(\mathfrak{m}) = N(\mathfrak{b})(\mathfrak{b}, \mathfrak{m}) = N(\mathfrak{b})(\mathfrak{b}, \mathfrak{a}),$$
$$N(\mathfrak{a}) = N(\mathfrak{d})(\mathfrak{d}, \mathfrak{a}) = N(\mathfrak{d})(\mathfrak{b}, \mathfrak{a}).$$

Consequently $(\mathfrak{b}, \mathfrak{a})$ is nonzero and

$$N(\mathfrak{a})N(\mathfrak{b}) = N(\mathfrak{m})N(\mathfrak{d}).$$

8. If the ideal $\mathfrak{a}$ is divisible by the ideal $\mathfrak{b}$ then, by §6, 2,

$$N(\mathfrak{a}) = (\mathfrak{b}, \mathfrak{a})N(\mathfrak{b}),$$

hence $N(\mathfrak{a})$ is divisible by $N(\mathfrak{b})$.

In particular, if $(\mathfrak{b}, \mathfrak{a}) = 1$, then $\mathfrak{b}$ is also divisible by $\mathfrak{a}$ and it follows that:

9. If $\mathfrak{a}$ is divisible by $\mathfrak{b}$ and at the same time $N(\mathfrak{a}) = N(\mathfrak{b})$, then $\mathfrak{a} = \mathfrak{b}$, i.e., the two ideals are identical.

10. If $\mathfrak{a}$ is divisible by $\mathfrak{a}_1$, and $\mathfrak{b}$ by $\mathfrak{b}_1$, then $\mathfrak{a}\mathfrak{b}$ is divisible by $\mathfrak{a}_1\mathfrak{b}_1$ (§4, 7).

11. If an ideal $\mathfrak{a}$ is divisible by a principal ideal $\mathfrak{o}\mu$, then all functions in $\mathfrak{a}$ are of the form $\beta\mu$, and the collection of all the functions β is again an ideal, $\mathfrak{b}$, so that one can set

$$\mathfrak{a} = \mu\mathfrak{b}.$$

12. If μ is an arbitrary nonzero function in $\mathfrak{o}$ and the ideal $\mathfrak{a}\mu$ is divisible by the ideal $\mathfrak{b}\mu$, then $\mathfrak{a}$ is divisible by $\mathfrak{b}$ and $\mathfrak{a}\mu = \mathfrak{b}\mu$ implies $\mathfrak{a} = \mathfrak{b}$.

13. The least common multiple of two ideals $\mathfrak{a}, \mathfrak{o}\nu$, one of which is a principal ideal, has the form $\mathfrak{r}\nu$ by 11, where $\mathfrak{r}$ is an ideal. Since, on the other hand, $\mathfrak{a}\nu$ is a common multiple of $\mathfrak{a}$ and $\mathfrak{o}\nu$, and hence divisible by $\mathfrak{r}\nu$, it follows by 12 that $\mathfrak{r}$ is a divisor of $\mathfrak{a}$.

14. If $\mathfrak{a}$ is an ideal and ν is a function in $\mathfrak{o}$ then, by §6, 2, 5:

$$(\mathfrak{o}, \mathfrak{a}\nu) = (\mathfrak{o}, \mathfrak{o}\nu)(\mathfrak{o}\nu, \mathfrak{a}\nu) = (\mathfrak{o}, \mathfrak{o}\nu)(\mathfrak{o}, \mathfrak{a}),$$

hence

$$N(\mathfrak{a}\nu) = \mathrm{const.}N(\mathfrak{a})N(\nu).$$

Thus if $\mathfrak{r}\nu$ is the least common multiple, and $\mathfrak{d}$ the greatest common divisor, of the two ideals $\mathfrak{a}, \mathfrak{o}\nu$, it follows from 7 that

$$N(\mathfrak{a}) = N(\mathfrak{r})N(\mathfrak{d}).$$

15. Each ideal $\mathfrak{a}$ different from $\mathfrak{o}$ is divisible by a prime ideal $\mathfrak{p}$.

If $\mathfrak{a}$ is not a prime ideal, then it has at least one strict divisor different from $\mathfrak{o}$, and among these let $\mathfrak{p}$ be one whose norm has minimal degree. The latter can have no strict divisor $\mathfrak{p}'$ different from $\mathfrak{o}$, otherwise $\mathfrak{p}'$ would also be a divisor of $\mathfrak{a}$ and (by 8) $N(\mathfrak{p}')$ would have lower degree than $N(\mathfrak{p})$. This contradicts the hypothesis on $\mathfrak{p}$, and hence $\mathfrak{p}$ is a prime ideal.

16. If $\mathfrak{a}$ is relatively prime to $\mathfrak{b}$ then $\mathfrak{a}\mathfrak{b}$ is the least common multiple of $\mathfrak{a}$ and $\mathfrak{b}$, and hence each ideal divisible by both $\mathfrak{a}$ and $\mathfrak{b}$ is also divisible by the product $\mathfrak{a}\mathfrak{b}$.

For, by hypothesis, there are functions α_1, β_1 in $\mathfrak{a}, \mathfrak{b}$ respectively such that

$$\alpha_1 + \beta_1 = 1,$$

by (5). On the other hand, if $\alpha = \beta$ is a function in the least common multiple $\mathfrak{m}$ of $\mathfrak{a}$ and $\mathfrak{b}$, then the above equation gives

$$\alpha = \beta = \alpha_1\beta + \alpha\beta_1,$$

showing it to be a function in $\mathfrak{a}\mathfrak{b}$. This means that $\mathfrak{m}$ is divisible by $\mathfrak{a}\mathfrak{b}$, and since, conversely, (by 2) $\mathfrak{a}\mathfrak{b}$ is divisible by $\mathfrak{m}$, $\mathfrak{m}$ is identical to $\mathfrak{a}\mathfrak{b}$, and it follows from 7 that in this case

$$N(\mathfrak{a}\mathfrak{b}) = N(\mathfrak{a})N(\mathfrak{b}).$$

17. If $\mathfrak{a}$ is an arbitrary ideal, and $\mathfrak{p}$ is a prime ideal, then either $\mathfrak{a}$ is divisible by $\mathfrak{p}$ or $\mathfrak{a}$ is relatively prime to $\mathfrak{p}$. Because, there being no divisors of $\mathfrak{p}$ other than $\mathfrak{o}$ and $\mathfrak{p}$, the greatest common divisor of $\mathfrak{a}$ and $\mathfrak{p}$ can be nothing but $\mathfrak{o}$ or $\mathfrak{p}$.

18. If $\mathfrak{a}$ is relatively prime to $\mathfrak{b}$ and to $\mathfrak{c}$, then $\mathfrak{a}$ is also relatively prime to $\mathfrak{b}\mathfrak{c}$.

By hypothesis there are (by 5) functions β, γ in $\mathfrak{b}, \mathfrak{c}$ satisfying the congruences

$$\beta \equiv 1, \quad \gamma \equiv 1 \ (\mathrm{mod} \ \mathfrak{a}),$$

hence by §7

$$\beta\gamma \equiv 1 \ (\mathrm{mod} \ \mathfrak{a}).$$

Since $\beta\gamma$ is in $\mathfrak{b}\mathfrak{c}$, this proves the assertion.

It also follows that, when the product $\mathfrak{a}\mathfrak{b}$ is divisible by a prime ideal $\mathfrak{p}$, at least one of the two factors $\mathfrak{a}, \mathfrak{b}$ must be divisible by $\mathfrak{p}$, and, applied to principal ideals, this means that when the product $\mu\nu$ of two integral algebraic functions is in $\mathfrak{p}$ at least one of the two factors μ, ν must be in $\mathfrak{p}$.

19. If $\mathfrak{a}$ is relatively prime to $\mathfrak{c}$ and $\mathfrak{ab}$ is divisible by $\mathfrak{c}$, then $\mathfrak{b}$ is divisible by $\mathfrak{c}$. By hypothesis there is a function α in $\mathfrak{a}$ satisfying the congruence

$$\alpha \equiv 1 \pmod{\mathfrak{c}}.$$

It follows that if β is any function in $\mathfrak{b}$ then

$$\beta \equiv \alpha\beta, \text{ which is } \equiv 0 \pmod{\mathfrak{c}}$$

by hypothesis. Consequently, β is in $\mathfrak{c}$, and hence $\mathfrak{b}$ is divisible by $\mathfrak{c}$.

§9. Laws of divisibility of ideals

SUMMARY AND COMMENTS

In ordinary arithmetic, the prime divisor property, which was proved by Euclid, is only a short step away from unique prime factorization. In algebraic number theory this is not the case (see Dedekind (1877), §§22, 23), because if $\mathfrak{p}$ divides $\mathfrak{a}$ it does not easily follow that $\mathfrak{a} = \mathfrak{pc}$ for some $\mathfrak{c}$. However, in algebraic function theory it *is* only a short step, carried out in this section.

The key fact, proved in subsection 4, is that if an ideal $\mathfrak{a}$ divides an ideal $\mathfrak{c}$ then $\mathfrak{c} = \mathfrak{ab}$ for some ideal $\mathfrak{b}$. This fact is obtainable more simply for algebraic functions than for algebraic numbers, thanks to some advantages that polynomial functions have over ordinary integers, notably their factorization into linear factors.

A useful consequence of unique prime factorization is the multiplicative property of the norm, $N(\mathfrak{ab}) = N(\mathfrak{a})N(\mathfrak{b})$, obtained in §8 only for relatively prime $\mathfrak{a}$ and $\mathfrak{b}$. The general case follows immediately from this special case once we have a unique prime ideal factorization.

Another important way in which the theory of prime ideals is simpler in the function field case is the following: the prime ideals are precisely those of first degree (see below, subsection 7). This is due to the fact that each polynomial splits into linear factors, thanks to the fundamental theorem of algebra.

All these theorems, which are mostly immediate from the definition of ideal, do not suffice to prove the complete analogy between the laws of divisibility of ideals and those of polynomial functions. Our proof of this will be based on the following theorem:

1. *If $\mathfrak{a}$ is an ideal and k is an arbitrary polynomial function of z, then a function α in $\mathfrak{a}$ may be chosen so that $(\mathfrak{a}, \mathfrak{o}\alpha)$ has no common divisor with k.*[24]
Namely, if

$$\mathfrak{a} = [\alpha_1, \alpha_2, \ldots, \alpha_n],$$
$$\mathfrak{o} = [\omega_1, \omega_2, \ldots, \omega_n],$$

---

[24]The fact that this theorem can be proved at this early stage distinguishes the theory of algebraic functions from that of algebraic numbers, and allows the former to proceed significantly more simply than the latter.

and if α is any function in $\mathfrak{a}$, then polynomial functions $x_{h,k}$ may be determined so that

$$\alpha\omega_1 = x_{1,1}\alpha_1 + x_{2,1}\alpha_2 + \cdots + x_{n,1}\alpha_n,$$
$$\alpha\omega_2 = x_{1,2}\alpha_1 + x_{2,2}\alpha_2 + \cdots + x_{n,2}\alpha_n,$$
$$\cdots\cdots\cdots\cdots\cdots\cdots\cdots\cdots\cdots\cdots\cdots\cdots$$
$$\alpha\omega_n = x_{1,n}\alpha_1 + x_{2,n}\alpha_2 + \cdots + x_{n,n}\alpha_n,$$

and (by §6, 4)

$$(\mathfrak{a}, \mathfrak{o}\alpha) = \mathrm{const.} \sum \pm x_{1,1}x_{2,2}\cdots x_{n,n}.$$

Now if $\sum \pm x_{1,1}x_{2,2}\cdots x_{n,n}$ is divisible by a linear factor $z-c$ of k, then it is possible to determine a function ω in $\mathfrak{o}$, not divisible by $z-c$, and a function α' in $\mathfrak{a}$ so that

$$\alpha\omega = (z-c)\alpha'.^{25}$$

If one now sets

$$t(z-c) - \omega = \omega',$$

where t is an undetermined constant, then

$$N(\omega') = t^n(z-c)^n + a_1 t^{n-1}(z-c)^{n-1} + \cdots + a_{n-1}t(z-c) + a_n,$$

where the coefficients $a_1, a_2, \ldots, a_n$ independent of t are polynomial functions of z.

It is now impossible for a_1 to be divisible by $z-c$, a_2 by $(z-c)^2$, ..., and a_n by $(z-c)^n$; otherwise $\frac{\omega}{z-c}$ would be an integral algebraic function (§2, 5; §3, 4), contrary to hypothesis. Therefore not all terms of $N(\omega')$ are divisible by $(z-c)^n$, and if $(z-c)^{n-r}$ is the highest power of $z-c$ that is so divisible, then $r > 0$ and

$$\frac{N(\omega')}{(z-c)^{n-r}} = t^n(z-c)^r + b_1 t^{n-1} + \cdots + b_{n-1}t + b_n = f(t),$$

where the polynomial functions $b_1, b_2, \ldots, b_n$ do not all vanish for $z = c$. Consequently there are only a finite number of constant values t for which $f(t)$ is divisible by $z-c$.

If $z-c'$ is a linear function of k different from $z-c$, then $f(t)$ is also divisible by $z-c'$ for only finitely many values of t. It follows that one can arrange the value of t so that $N(\omega')$ is not divisible by $(z-c)^n$ or any other linear factor of k.[26] When this is done, if one sets

$$t\alpha - \alpha' = \alpha'',$$

which is likewise a function in $\mathfrak{a}$, then it follows that

$$\alpha\omega' = (z-c)\alpha'',$$
$$N(\alpha'') = \frac{N(\alpha)N(\omega')}{(z-c)^n},$$

---

[25]Namely, if the determinant $\sum \pm x_{1,1}x_{2,2}\cdots x_{n,n}$ is divisible by $z-c$, i.e., if it vanishes for $z = c$, then one can determine a system of constants $c_1, c_2, \ldots, c_n$, not all zero, so that the polynomial functions

$$c_1 x_{k,1} + c_2 x_{k,2} + \cdots + c_k x_{k,n} \qquad (k = 1, 2, \ldots, n)$$

vanish for $z = c$, and hence are divisible by $z-c$, and one then sets

$$\omega = c_1\omega_1 + c_2\omega_2 + \cdots + c_n\omega_n.$$

[26]This conclusion is not valid for the analogous question in number theory.

and therefore, since

$$(\mathfrak{a}, \mathfrak{o}\alpha) = \text{const.} \frac{N(\alpha)}{N(\mathfrak{a})}$$

by §7, (4), we have:

$$(\mathfrak{a}, \mathfrak{o}\alpha'') = \text{const.} \frac{(\mathfrak{a}, \mathfrak{o}\alpha)N(\omega')}{(z - c)^n}.$$

Thus the function $(\mathfrak{a}, \mathfrak{o}\alpha'')$ contains the factor $z - c$ at least once less often than does $(\mathfrak{a}, \mathfrak{o}\alpha)$, which at the same time can contain no other linear factor of k more often than does $(\mathfrak{a}, \mathfrak{o}\alpha)$. The theorem therefore follows by repeated application of this process.

2. Each ideal $\mathfrak{a}$ can be expressed as the greatest common divisor of two principal ideals $\mathfrak{o}\mu$, $\mathfrak{o}\nu$, one of which can be taken as an arbitrary ideal divisible by $\mathfrak{a}$.

Proof. One chooses an arbitrary nonzero function ν in $\mathfrak{a}$ and (by 1) a second function μ such that the two functions $(\mathfrak{a}, \mathfrak{o}\nu)$ and $(\mathfrak{a}, \mathfrak{o}\mu)$ have no common divisor. Now if α is any function in $\mathfrak{a}$ then it follows by §6 that $(\mathfrak{a}, \mathfrak{o}\mu)\alpha$ is in $\mathfrak{o}\mu$ and $(\mathfrak{a}, \mathfrak{o}\nu)\alpha$ is in $\mathfrak{o}\nu$, so there are two functions ω, ω' in $\mathfrak{o}$ for which

$$(\mathfrak{a}, \mathfrak{o}\mu)\alpha = \mu\omega, \quad (\mathfrak{a}, \mathfrak{o}\nu)\alpha = \nu\omega'.$$

Hence if one chooses two polynomial functions g, h of z satisfying

$$g(\mathfrak{a}, \mathfrak{o}\mu) + h(\mathfrak{a}, \mathfrak{o}\nu) = 1,$$

as is possible by the hypothesis on $(\mathfrak{a}, \mathfrak{o}\mu)$ and $(\mathfrak{a}, \mathfrak{o}\nu)$, then it follows that

$$\alpha = g\mu\omega + h\nu\omega',$$

i.e., $\mathfrak{a}$ is divisible by the greatest common divisor of $\mathfrak{o}\mu$ and $\mathfrak{o}\nu$. And since the latter, conversely, is divisible by $\mathfrak{a}$ (since $\mathfrak{o}\mu$ and $\mathfrak{o}\nu$ are divisible by $\mathfrak{a}$), it is equal to $\mathfrak{a}$.

Q.E.D.

3. Each ideal $\mathfrak{a}$ can be converted into a principal ideal $\mathfrak{o}\mu = \mathfrak{a}\mathfrak{m}$ by multiplying by an ideal $\mathfrak{m}$.[27]

Proof. By 1, one chooses a function μ in $\mathfrak{a}$ such that $(\mathfrak{a}, \mathfrak{o}\mu)$ has no divisor in common with $N(\mathfrak{a})$, and then a second function ν such that $(\mathfrak{a}, \mathfrak{o}\nu)$ has no divisor in common with $(\mathfrak{a}, \mathfrak{o}\mu)$. Then, by 2, $\mathfrak{a}$ is the greatest common divisor of $\mathfrak{o}\mu$ and $\mathfrak{o}\nu$. By §8, 13, the least common multiple of $\mathfrak{o}\mu$ and $\mathfrak{o}\nu$ is of the form $\mathfrak{m}\nu$, where $\mathfrak{m}$ is a divisor of $\mathfrak{o}\mu$. Then by §8, 14

$$N(\mathfrak{m}) = \frac{N(\mathfrak{o}\mu)}{N(\mathfrak{a})} = (\mathfrak{a}, \mathfrak{o}\mu),$$

hence, by hypothesis, it has no divisor in common with $N(\mathfrak{a})$. Thus if one again determines two polynomial functions g, h of z so that

$$gN(\mathfrak{m}) + hN(\mathfrak{a}) = 1,$$

then it follows by §8, 5, since $N(\mathfrak{m})$ is in $\mathfrak{m}$ and $N(\mathfrak{a})$ is in $\mathfrak{a}$, that $\mathfrak{m}$ and $\mathfrak{a}$ are relatively prime ideals. Hence, by §8, 16,

$$N(\mathfrak{m}\mathfrak{a}) = N(\mathfrak{m})N(\mathfrak{a}) = N(\mathfrak{o}\mu).$$

---

[27]One can choose the ideal $\mathfrak{m}$ so that it is also relatively prime to an arbitrary ideal $\mathfrak{b}$. This is achieved by taking the function μ so that $(\mathfrak{a}, \mathfrak{o}\mu) = N(\mathfrak{m})$ has no divisor in common with $N(\mathfrak{a})N(\mathfrak{b})$ (§8, 8).

Now since $\mathfrak{o}\mu$ is divisible by $\mathfrak{m}$ and $\mathfrak{a}$, and hence also by $\mathfrak{m}\mathfrak{a}$ (§8, 16), it follows by §8, 9 that

$$\mathfrak{m}\mathfrak{a} = \mathfrak{o}\mu.$$

Q.E.D.

4. If an ideal $\mathfrak{c}$ is divisible by an ideal $\mathfrak{a}$, then there is one and *only* one ideal $\mathfrak{b}$ that satisfies the condition

$$\mathfrak{a}\mathfrak{b} = \mathfrak{c}.$$

It is called the *quotient of $\mathfrak{c}$ by $\mathfrak{a}$*.[28]

If $\mathfrak{a}\mathfrak{b}$ is divisible by $\mathfrak{a}\mathfrak{b}'$ then $\mathfrak{b}$ is divisible by $\mathfrak{b}'$ and $\mathfrak{a}\mathfrak{b} = \mathfrak{a}\mathfrak{b}'$ implies $\mathfrak{b} = \mathfrak{b}'$.

Proof. Suppose $\mathfrak{c}$ is divisible by $\mathfrak{a}$ and suppose (by 3) that $\mathfrak{a}\mathfrak{m} = \mathfrak{o}\mu$. Then $\mathfrak{c}\mathfrak{m}$ is also divisible by $\mathfrak{a}\mathfrak{m} = \mathfrak{o}\mu$ and consequently $\mathfrak{c}\mathfrak{m} = \mathfrak{b}\mu$ (§8, 10, 11). Thus, multiplying the last equation by $\mathfrak{a}$,

$$\mathfrak{c}\mu = \mathfrak{a}\mathfrak{b}\mu,$$

and by §8, 12

$$\mathfrak{c} = \mathfrak{a}\mathfrak{b},$$

proving the first part of the theorem.[29]

If also $\mathfrak{a}\mathfrak{b}$ is divisible by $\mathfrak{a}\mathfrak{b}'$ then (§8, 10) $\mu\mathfrak{b}$ is divisible by $\mu\mathfrak{b}$, hence $\mathfrak{b}$ is divisible by $\mathfrak{b}'$. If $\mathfrak{a}\mathfrak{b} = \mathfrak{a}\mathfrak{b}'$ then it follows that $\mu\mathfrak{b} = \mu\mathfrak{b}'$, and consequently $\mathfrak{b} = \mathfrak{b}'$.

5. Each ideal different from $\mathfrak{o}$ is either a prime ideal, or else uniquely expressible as a product of prime ideals.

Proof. If the ideal $\mathfrak{a}$ is different from $\mathfrak{o}$ then (by §8, 15) it is divisible by a prime ideal $\mathfrak{p}_1$ and hence (by 4) $\mathfrak{a} = \mathfrak{p}_1\mathfrak{a}_1$, where $\mathfrak{a}_1$ is a *strict* divisor of $\mathfrak{a}$ (because $\mathfrak{a}_1 = \mathfrak{a}$ implies $\mathfrak{p}_1 = \mathfrak{o}$ by 4). Thus the degree of $N(\mathfrak{a}_1)$ is lower than that of $N(\mathfrak{a})$. If $\mathfrak{a}_1$ is different from $\mathfrak{o}$, then one concludes similarly that $\mathfrak{a}_1 = \mathfrak{p}_2\mathfrak{a}_2$, where the degree of $N(\mathfrak{a}_2)$ is lower than that of $N(\mathfrak{a}_1)$. Continuing in this way, after a finite number of steps one arrives at an ideal $\mathfrak{a}_{r-1} = \mathfrak{p}_r\mathfrak{a}_r$ with $N(\mathfrak{a}_r) = 1$, so $\mathfrak{a}_r = \mathfrak{o}$. Therefore

$$\mathfrak{a} = \mathfrak{p}_1\mathfrak{p}_2 \cdots \mathfrak{p}_r.$$

If there were a second such decomposition, say

$$\mathfrak{p}_1\mathfrak{p}_2 \cdots \mathfrak{p}_r = \mathfrak{q}_1\mathfrak{q}_2 \cdots \mathfrak{q}_s,$$

then (by §8, 18) at least one of the prime ideals $\mathfrak{p}_1, \mathfrak{p}_2, \ldots, \mathfrak{p}_r$, say $\mathfrak{p}_1$, must be divisible by $\mathfrak{q}_1$ and hence equal to $\mathfrak{q}_1$. Then, by 4,

$$\mathfrak{p}_2\mathfrak{p}_3 \cdots \mathfrak{p}_r = \mathfrak{q}_2\mathfrak{q}_3 \cdots \mathfrak{q}_s.$$

From this one concludes similarly that $\mathfrak{p}_2 = \mathfrak{q}_2$, etc.

Collecting equal primes in the decomposition just obtained, one can set

$$\mathfrak{a} = \mathfrak{p}_1^{e_1}\mathfrak{p}_2^{e_2} \cdots \mathfrak{p}_r^{e_r}.$$

Then a divisor $\mathfrak{a}_1$ of $\mathfrak{a}$ can be divisible by no prime ideals except $\mathfrak{p}_1, \mathfrak{p}_2, \ldots, \mathfrak{p}_r$, and no more often than $\mathfrak{a}$ is. Thus one obtains all the divisors of $\mathfrak{a}$, the number of which is finite and equal to $(e_1 + 1)(e_2 + 1) \cdots (e_r + 1)$, by letting the exponents h_ι in

$$\mathfrak{p}_1^{h_1}\mathfrak{p}_2^{h_2} \cdots \mathfrak{p}_r^{h_r}$$

---

[28]Recall that this was the particular difficulty in algebraic number theory pointed by Dedekind (1877). (Translator's note.)

[29]This definition of quotient of two ideals agrees completely with that given in §4, 8.

run through the numbers $0, 1, 2, \ldots, e_\iota$ (where $\mathfrak{p}^0$ is understood to be the ideal $\mathfrak{o}$). If $\mathfrak{a}, \mathfrak{b}$ are two ideals

$$\mathfrak{a} = \mathfrak{p}_1^{e_1} \mathfrak{p}_2^{e_2} \cdots \mathfrak{p}_r^{e_r}; \quad \mathfrak{b} = \mathfrak{p}_1^{f_1} \mathfrak{p}_2^{f_2} \cdots \mathfrak{p}_r^{f_r}$$

(where some of the exponents e, f can be zero), then one obtains the greatest common divisor and the least common multiple of $\mathfrak{a}$ and $\mathfrak{b}$ in the form

$$\mathfrak{p}_1^{g_1} \mathfrak{p}_2^{g_2} \cdots \mathfrak{p}_r^{g_r}$$

by taking $g_1, g_2, \ldots, g_r$ to be the least members of the respective pairs e_1, f_1; e_2, f_2; $\ldots$; e_r, f_r in the former case, and the greatest in the latter case.

6. If $\mathfrak{a}, \mathfrak{b}$ are any two ideals, then

$$N(\mathfrak{a}\mathfrak{b}) = N(\mathfrak{a})N(\mathfrak{b}).$$

Proof. Suppose, as in 5, that $\mathfrak{a} = \mathfrak{p}_1 \mathfrak{a}_1$. Then, since $\mathfrak{a}_1$ is a strict divisor of $\mathfrak{a}$, there is a function η in $\mathfrak{a}_1$ not divisible by $\mathfrak{a}$. The least common multiple and the greatest common divisor of $\mathfrak{a}$ and $\mathfrak{o}\eta$ are $\mathfrak{p}_1\eta$ and $\mathfrak{a}_1$ respectively, as follows immediately (by 5) from the decomposition of $\mathfrak{a}$ and $\mathfrak{o}\mu$ into their prime factors. But then it follows from §8, 14 that

$$N(\mathfrak{a}) = N(\mathfrak{p}_1)N(\mathfrak{a}_1).$$

Repeating this argument for $\mathfrak{a}_1$ etc., we get

$$N(\mathfrak{a}) = N(\mathfrak{p}_1)N(\mathfrak{p}_2) \cdots N(\mathfrak{p}_r),$$

when $\mathfrak{a} = \mathfrak{p}_1 \mathfrak{p}_2 \cdots \mathfrak{p}_r$, and this gives

$$N(\mathfrak{a}\mathfrak{b}) = N(\mathfrak{a})N(\mathfrak{b}).$$

7. Each prime ideal is an ideal of *first degree* (§7) and, conversely, each ideal of first degree is a prime ideal.[30]

Proof. If $\mathfrak{p}$ is a prime ideal, then $N(\mathfrak{p})$ is divisible by $\mathfrak{p}$, and hence at least one of the linear factors of $N(\mathfrak{p})$, say $z - c$, is divisible by $\mathfrak{p}$ (§8, 18). If ω is an arbitrary function in $\mathfrak{o}$, satisfying the equation:

$$\omega^n + a_1 \omega^{n-1} + \cdots + a_{n-1}\omega + a_n = 0,$$

then, by replacing the polynomial functions $a_1, a_2, \ldots, a_n$ by their constant remainders $a_1^{(0)}, a_2^{(0)}, \ldots, a_n^{(0)}$ modulo $z - c$, one obtains the integral algebraic function

$$\omega^n + a_1^{(0)} \omega^{n-1} + \cdots + a_{n-1}^{(0)}\omega + a_n^{(0)},$$

and decomposing this into linear factors $(\omega - b_1), (\omega - b_2), \ldots, (\omega - b_n)$:

$$(\omega - b_1)(\omega - b_2) \cdots (\omega - b_n) = (z - c)\omega' \equiv 0 \pmod{\mathfrak{p}}.$$

Thus at least one of the factors $\omega - b_1, \omega - b_2, \ldots$ must be divisible by $\mathfrak{p}$, i.e.,

$$\omega \equiv b \pmod{\mathfrak{p}},$$

where b is a constant. Since this means every function in $\mathfrak{o}$ is congruent to a constant $\pmod{\mathfrak{p}}$, it follows by §6 that $(\mathfrak{o}, \mathfrak{p}) = N(\mathfrak{p}) = z - c$ is a *linear* function of z, and the first part of the assertion is proved.

Conversely: if $\mathfrak{q}$ is an ideal of first degree, and

$$N(\mathfrak{q}) = z - c,$$

---

[30]This theorem distinguishes the theory of algebraic functions from the analogous theory of algebraic numbers.

then $\mathfrak{q}$ is certainly divisible by a prime ideal $\mathfrak{p}$, and since $N(\mathfrak{q})$ is divisible by $N(\mathfrak{p})$, $N(\mathfrak{p}) = N(\mathfrak{q}) = z - c$. Hence, by §8, 9,

$$\mathfrak{p} = \mathfrak{q}.$$

It follows from this that the degree of an ideal is equal to the number of factors in its prime decomposition. Therefore if

$$\mathfrak{o}(z - c) = \mathfrak{p}_1^{e_1} \mathfrak{p}_2^{e_2} \mathfrak{p}_3^{e_3} \cdots ,$$

then

$$e_1 + e_2 + e_3 + \cdots = n.$$

It also follows that a *polynomial* function of z is divisible by a prime ideal $\mathfrak{p}$ if and only if it is divisible by the norm of $\mathfrak{p}$.

§10. Complementary bases of the field Ω

Summary and Comments

For each basis $\alpha_1, \ldots, \alpha_n$ of Ω there is a basis $\alpha_1', \ldots, \alpha_n'$, called the *complementary basis*. It is defined with the help of the trace and has discriminant

$$\Delta(\alpha_1', \ldots, \alpha_n') = 1/\Delta(\alpha_1, \ldots, \alpha_n).$$

When $\alpha_1, \ldots, \alpha_n$ is the basis of a module $\mathfrak{a}$, the complementary basis is the basis of a *complementary module* $\mathfrak{a}'$, which is independent of the basis chosen for $\mathfrak{a}$.

In particular, when $\mathfrak{e}$ is the complementary module of $\mathfrak{o}$, and D is the discriminant of Ω, then $D\mathfrak{e}$ is an ideal. The module $\mathfrak{e}$ is related to ramification in §11, and it comes up again in §23, where it is shown that if $\omega \in \mathfrak{o}$ then $\frac{d\omega}{dz} \in \mathfrak{e}$. Finally, in §32, $\mathfrak{e}$ is used to conclude the proof of the residue theorem. The ideal $\mathfrak{f}$, which follows on the heels of $\mathfrak{e}$ in this section and the next, also reappears in §23 as the "ideal of the double points."

The general concept of complementary module (more precisely, the related concept of *supplementary polygon*) has a key role to play in the proof of the Riemann-Roch theorem, in §27. The concept of complementary basis is also involved in the proof of Abel's theorem, §26.

Complementary modules can be viewed as the module counterpart of dual vector spaces, and they are introduced as such in Eichler (1966), where their role is highlighted in a "Riemann-Roch theorem for linear divisors" on p. 24. However, it is still a long road to the classical Riemann-Roch theorem. For Dedekind and Weber the crucial extra ingredient is the concept of differential, introduced in §23. Already in subsection 8 of this section one sees a relationship between the concepts of complementary basis and the derivative of a polynomial.

1. *Definition.* If the functions $\alpha_1, \alpha_2, \ldots, \alpha_n$ form a basis of Ω, and if one makes the abbreviations

$$Tr(\alpha_r \alpha_s) = a_{r,s} = a_{s,r},$$

$$\Delta(\alpha_1, \alpha_2, \ldots, \alpha_n) = \sum \pm a_{1,1} a_{2,2} \ldots a_{n,n} = a \qquad (\S 2),$$

then, since a is nonzero, it is possible to determine a system of functions $\alpha_1', \alpha_2', \ldots, \alpha_n'$ by the linear equations

(1)
$$\alpha_r = \sum_{\iota=1}^{n} a_{r,\iota} \alpha_\iota',$$

and since

$$\Delta(\alpha_1', \alpha_2', \ldots, \alpha_n') = \frac{1}{a}$$

is nonzero, the functions $\alpha_1', \alpha_2', \ldots, \alpha_n'$ likewise form a basis of Ω. It will be called the *complementary basis* of $\alpha_1, \alpha_2, \ldots, \alpha_n$.

2. If δ_{rs} denotes the number 1 or 0 according as r, s are equal or not,[31] where the indices r, s belong to the series $1, 2, \ldots, n$, then

(2)
$$Tr(\alpha_r \alpha_s') = \delta_{rs},$$

because it follows by solution of the equations (1) that

$$\alpha_s' = \sum_\iota a_{\iota,s}' \alpha_\iota;$$

$$a_{r,s}' = a_{s,r}'; \quad \sum_\iota a_{r,\iota} a_{s,\iota}' = \delta_{rs},$$

whence:

$$\alpha_r \alpha_s' = \sum_\iota a_{\iota,s}' \alpha_\iota \alpha_r; \quad Tr(\alpha_r \alpha_s') = \sum_\iota a_{\iota,s}' a_{\iota,r} = \delta_{rs}.$$

Conversely, if a function system β_s satisfies the conditions $Tr(\alpha_r \beta_s) = \delta_{rs}$, then $\beta_s = \alpha_s'$. Because, if one sets $\beta_s = \sum_\iota b_{\iota,s} \alpha_\iota'$, then it follows from (2) that

$$b_{r,s} = Tr(\beta_s \alpha_r) = \delta_{rs}.$$

This implies that the relation between the α_ι and the α_ι' is symmetric, i.e., that the basis $\alpha_1, \alpha_2, \ldots, \alpha_n$ is complementary to $\alpha_1', \alpha_2', \ldots, \alpha_n'$.

3. If η is an arbitrary function in Ω, then one can always set

$$\eta = \sum x_\iota \alpha_\iota = \sum x_\iota' \alpha_\iota',$$

and it follows by application of (2) that

$$x_\iota = Tr(\eta \alpha_\iota'), \quad x' = Tr(\eta \alpha_\iota),$$

hence:

(3)
$$\eta = \sum_\iota \alpha_\iota Tr(\eta \alpha_\iota') = \sum_\iota \alpha_\iota' Tr(\eta \alpha_\iota).$$

4. If η is any nonzero function in Ω, then

$$\frac{\alpha_1'}{\eta}, \quad \frac{\alpha_2'}{\eta}, \quad \ldots, \quad \frac{\alpha_n'}{\eta}$$

---

[31]Dedekind and Weber denote this number by (r, s). But since this notation is already overused, I have changed it to the now-familiar Kronecker delta notation. (Translator's note.)

is the basis complementary to $\eta\alpha_1, \eta\alpha_2, \ldots, \eta\alpha_n$. This follows from 2 because

$$Tr\left(\eta\alpha_r \cdot \frac{\alpha'_s}{\eta}\right) = Tr(\alpha_r \alpha'_s) = \delta_{rs}.$$

5. When two bases of Ω, $\alpha_1, \alpha_2, \ldots, \alpha_n$ and $\beta_1, \beta_2, \ldots, \beta_n$, are connected by the n equations

$$\beta_s = \sum_\iota x_{\iota,s}\alpha_\iota$$

with rational coefficients $x_{\iota,s}$, their complementary bases are connected by the n equations

$$\alpha'_r = \sum_\iota x_{r,\iota}\beta'_\iota$$

(transposed substitution). This is an immediate consequence of 3 because

$$x_{r,s} = Tr(\alpha'_r\beta_s).$$

6. One has

$$\sum_\iota \alpha_\iota \alpha'_\iota = 1,$$

hence:

$$\sum_{\iota,\iota'} a_{\iota,\iota'}\alpha'_\iota\alpha'_{\iota'} = \sum_{\iota,\iota'} a'_{\iota,\iota'}\alpha_\iota\alpha_{\iota'} = 1.$$

Namely, if one first sets

$$\sum_\iota \alpha_\iota\alpha_{\iota'} = \sigma,$$

then it follows from 3 (applied to the functions $\eta\alpha_r$) that

$$\eta\alpha_r = \sum_\iota \alpha_\iota Tr(\eta\alpha_r\alpha'_\iota).$$

Hence, by definition of the trace in §2, (5),

$$Tr(\eta) = \sum_\iota Tr(\eta\alpha_\iota\alpha'_\iota) = Tr\left(\sum \eta\alpha_\iota\alpha'_\iota\right).$$

So

$$Tr(\eta\sigma) = Tr(\eta),$$

and when one substitutes $\eta = \sigma$ and then $\eta = 1$ in 3,

$$\sigma = \sum_\iota \alpha_\iota Tr(\sigma\alpha'_1) = \sum_\iota \alpha_\iota Tr(\alpha'_\iota) = 1.$$

We now look in more detail at the construction of the complementary bases in two special cases:

7. Let $\omega_1, \omega_2, \ldots, \omega_n$ be a basis of $\mathfrak{o}$ and let $\varepsilon_1, \varepsilon_1, \ldots, \varepsilon_n$ be the complementary basis (of Ω). Let

$$e_{r,s} = e_{s,r} = Tr(\omega_r\omega_s),$$

which are *polynomial* functions, and let

$$D = \text{const.} \sum \pm e_{1,1}e_{2,2}\cdots e_{n,n}$$

be the discriminant of Ω. Then by 2

$$\varepsilon_r = \frac{1}{D}\sum_\iota \frac{\partial D}{\partial e_{\iota,r}}\omega_\iota,$$

whence it follows that the functions $D\varepsilon_r$ are *integral algebraic functions*. But it also follows from 6 that

$$D = \sum_{\iota,\iota'} \frac{\partial D}{\partial e_{\iota,\iota'}} \omega_\iota \omega_{\iota'},$$

whence

$$\varepsilon_r \varepsilon_s = \frac{1}{D^2} \sum_{\iota,\iota'} \frac{\partial D}{\partial e_{r,\iota}} \frac{\partial D}{\partial e_{s,\iota'}} \omega_\iota \omega_{\iota'}$$

$$= \frac{1}{D} \frac{\partial D}{\partial e_{r,s}} + \frac{1}{D^2} \sum_{\iota,\iota'} \left(\frac{\partial D}{\partial e_{r,\iota}} \frac{\partial D}{\partial e_{s,\iota'}} - \frac{\partial D}{\partial e_{r,s}} \frac{\partial D}{\partial e_{\iota,\iota'}} \right) \omega_\iota \omega_{\iota'},$$

and then it follows, by a well-known determinant theorem, that

$$\varepsilon_r \varepsilon_s = \frac{1}{D} \frac{\partial D}{\partial e_{r,s}} + \frac{1}{D} \sum_{\iota,\iota'} \frac{\partial^2 D}{\partial e_{r,\iota} \partial e_{s,\iota'}} \omega_\iota \omega_{\iota'},$$

which yields the important result that the functions $D\varepsilon_r \varepsilon_s$ are also *integral algebraic functions*.

8. Let θ be a function in Ω such that $1, \theta, \theta^2, \ldots, \theta^{n-1}$ form a basis of Ω, and suppose

(4) $$f(\theta) = \theta^n + a_1 \theta^{n-1} + \cdots + a_{n-1} \theta + a_n = 0$$

is irreducible with rational coefficients a. We seek the complementary basis to $1, \theta, \theta^2, \ldots, \theta^{n-1}$. For an undetermined constant t we set

$$\frac{f(t)}{t - \theta} = \eta_0 + \eta_1 t + \eta_2 t^2 + \cdots + \eta_{n-1} t^{n-1},$$

hence

(5) $$\begin{cases} \eta_0 = a_{n-1} + a_{n-2}\theta + \cdots + a_1 \theta^{n-2} + \theta^{n-1}, \\ \eta_1 = a_{n-2} + a_{n-3}\theta + \cdots + \theta^{n-2}, \\ \cdots\cdots\cdots\cdots\cdots\cdots\cdots \\ \eta_{n-2} = a_1 + \theta, \\ \eta_{n-1} = 1 \end{cases}$$

and so the functions $\eta_0, \eta_1, \ldots, \eta_{n-1}$ likewise form a basis of Ω, since the determinant of the equations (5) equals $(-1)^{\frac{1}{2}n(n-1)}$, and hence is nonzero. Each function ζ may therefore be expressed in the form

$$\zeta = y_0 \eta_0 + y_1 \eta_1 + \cdots + y_{n-1} \eta_{n-1}.$$

We now continue the series of rational functions $y_0, y_1, \ldots, y_{n-1}$ by determining functions $y_n, y_{n+1}, \ldots$ by the recursion

(6) $$a_n y_r + a_{n-1} y_{r+1} + \cdots + a_2 y_{r+n-2} + a_1 y_{r+n-1} + y_{r+n} = 0.$$

Now, by (5)

(7) $$\begin{cases} \theta \eta_0 = \qquad -a_n \eta_{n-1}, \\ \theta \eta_1 = \eta_0 - a_{n-1} \eta_{n-1}, \\ \theta \eta_2 = \eta_1 - a_{n-2} \eta_{n-1}, \\ \cdots\cdots\cdots\cdots\cdots\cdots\cdots \\ \theta \eta_{n-1} = \eta_{n-2} - a_1 \eta_{n-1}, \end{cases}$$

hence

$$\zeta\theta = y_1\eta_0 + y_2\eta_1 + \cdots + y_{n-1}\eta_{n-2} + y_n\eta_{n-1},$$

and similarly for each positive integer r:

$$\zeta\theta^r = y_r\eta_0 + y_{r+1}\eta_1 + \cdots + y_{r+n-2}\eta_{n-2} + y_{r+n-1}\eta_{n-1},$$

or, when one expresses $\eta_0, \eta_1, \ldots, \eta_{n-1}$ in terms of $1, \theta, \theta^2, \ldots, \theta^{n-1}$:

$$\zeta\theta^r = x_0^{(r)} + x_1^{(r)}\theta + x_2^{(r)}\theta^2 + \cdots + x_{n-1}^{(r)}\theta^{n-1},$$

where

$$x_0^{(r)} = y_r a_{n-1} + y_{r+1}a_{n-2} + \cdots + y_{r+n-2}a_1 + y_{r+n-1},$$

$$x_1^{(r)} = y_r a_{n-2} + y_{r+1}a_{n-3} + \cdots + y_{r+n-2},$$

$$\cdots\cdots\cdots\cdots\cdots\cdots\cdots\cdots$$

$$x_{n-2}^{(r)} = y_r a_1 + y_{r+1},$$

$$x_{n-1}^{(r)} = y_r.$$

Hence (by the definition of Tr, §2, (5)),

$$Tr(\zeta) = x_0^{(0)} + x_1^{(1)} + x_2^{(2)} + \cdots + x_{n-1}^{(n-1)}$$
$$= y_0 a_{n-1} + 2y_1 a_{n-2} + \cdots + (n-1)y_{n-2}a_1 + ny_{n-1}.$$

Thus, applying this to $\zeta = \eta_r$:

$$Tr(\eta_r) = (r+1)a_{n-1-r}; \quad Tr(\eta_{n-1-r}) = (n-r)a_r,$$

where a_0 is set equal to 1.

Hence if one makes the abbreviation[32]

$$Tr(\theta^r) = s_r,$$

then it follows from (5), for $r \leq n$, that

(8) $$(n-r)a_r = a_r s_0 + a_{r-1}s_1 + \cdots + a_1 s_{r-1} + s_r$$

and from (4) that in general

(9) $$0 = a_n s_r + a_{n-1}s_{r+1} + \cdots + a_1 s_{r+n-1} + s_{r+n}.$$

But these formulae also imply

(10) $$\begin{cases} f'(\theta) &= n\theta^{n-1} + (n-1)a_1\theta^{n-1} + \cdots + 2a_{n-2}\theta + a_{n-1} \\ &= s_0\eta_0 + s_1\eta_1 + \cdots + s_{n-1}\eta_{n-1}, \\ \theta^r f'(\theta) &= s_r\eta_0 + s_{r+1}\eta_1 + \cdots + s_{r+n-2}\eta_{n-2} + s_{r+n-1}\eta_{n-1}. \end{cases}$$

Bearing in mind the value of the determinant for the equation system (5), and the definitions of norm and discriminant in §2, (4) and (12), we now get the important

---

[32]Remember that Dedekind and Weber use S for trace ($=$"Spur" in German); this accounts for the notational choice of lower case s. (Translator's note.)

formula

$$(11) \qquad N f'(\theta) = (-1)^{\frac{1}{2}n(n-1)} \begin{vmatrix} s_0 & s_1 & \cdots & s_{n-1} \\ s_1 & s_2 & \cdots & s_n \\ \cdots & \cdots & \cdots & \cdots \\ s_{n-1} & s_n & \cdots & s_{2n-2} \end{vmatrix}$$

$$= (-1)^{\frac{1}{2}n(n-1)} \Delta(1, \theta, \theta^2, \ldots, \theta^{n-1}).$$

Looking back to Definition 1, we see that the equations (10) also yield the complementary basis of $1, \theta, \theta^2, \ldots, \theta^{n-1}$:

$$\frac{\eta_0}{f'(\theta)}, \quad \frac{\eta_1}{f'(\theta)}, \quad \cdots, \quad \frac{\eta_{n-1}}{f'(\theta)}.$$

9. If $\mathfrak{a} = [\alpha_1, \alpha_2, \ldots, \alpha_n]$ denotes a module whose basis is also a basis of Ω, then the basis $\alpha_1', \alpha_2', \ldots, \alpha_n'$ of Ω complementary to $\alpha_1, \alpha_2, \ldots, \alpha_n$ gives another module $\mathfrak{a}' = [\alpha_1', \alpha_2', \ldots, \alpha_n']$ called the *complementary module* of $\mathfrak{a}$. The latter, as follows immediately from 5, together with §4, 2, is independent of the choice of basis for $\mathfrak{a}$.

10. We consider in particular the module $\mathfrak{e} = [\varepsilon_1, \varepsilon_2, \ldots, \varepsilon_n]$ complementary to $\mathfrak{o} = [\omega_1, \omega_2, \ldots, \omega_n]$. If we set

$$\omega_r \omega_s = \sum_\iota e_{r,s}^{(\iota)} \omega_s,$$

then by 3

$$e_{r,s}^{(\iota)} = e_{s,r}^{(\iota)} = Tr(\omega_r \omega_s \varepsilon_\iota)$$

is a polynomial function of z, and it follows that

$$\omega_r \varepsilon_s = \sum_\iota e_{r,\iota}^{(s)} \varepsilon_\iota.$$

It follows from this that the module $\mathfrak{oe}$ (§4, 7) is divisible by $\mathfrak{e}$. On the other hand, since $\mathfrak{o}$ contains the function 1, $\mathfrak{e}$ is divisible by $\mathfrak{oe}$, hence

$$\mathfrak{oe} = \mathfrak{e}.$$

I.e., the module $\mathfrak{e}$, although it is not composed completely of integral algebraic functions, has the characteristic property II, §7, of ideals. Consequently, the same is true of the module $\mathfrak{e}^2$. Since the two modules $D\mathfrak{e}, D\mathfrak{e}^2$ contain only integral algebraic functions, by 7, they are both ideals, and it follows from 7 that

$$N(D\mathfrak{e}) = D^{n-1}.$$

11. If θ is a function in $\mathfrak{o}$ with the property that $1, \theta, \theta^2, \ldots, \theta^{n-1}$ forms a basis of Ω, so that the coefficients in the irreducible equation

$$f(\theta) = \theta^n + a_1 \theta^{n-1} + \cdots + a_{n-1}\theta + a_n = 0$$

are *polynomial* functions of z, then for $r = 0, 1, 2, \ldots, n-1$ one can determine the polynomial functions $k_\iota^{(r)}$ such that

$$\theta^r = \sum_{\iota=1}^n k_\iota^{(r)} \omega_\iota.$$

Applying Theorems 5 and 8 to this, one gets

$$f'(\theta)\varepsilon_s = k_s^{(0)}\eta_0 + k_s^{(1)}\eta_1 + \cdots + k_s^{(n-1)}\eta_{n-1},$$

which implies that the module

$$f'(\theta)\mathfrak{e} = \mathfrak{f}$$

contains only *integral algebraic* functions. One concludes from 10 that the latter is an *ideal*.

§11. The ramification ideal

Summary and Comments

The ramification ideal, as its name suggests, is associated with the ramification points of a Riemann surface. Unfortunately, "points" will not be introduced until §14, and "ramification points" only in §16, so the meaning of ramification ideal will probably be unclear until we arrive at §16.

Nevertheless, a glimpse of ramification may be caught near the end of subsection 2, where it is mentioned that "there are only a finite number of linear functions $z - c$ divisible by the square of a prime ideal." The corresponding finitely many values of c correspond to ramification points. As mentioned in Section 4 of the Translator's Introduction, ramification is crucial for the definition of genus in the Dedekind-Weber concept of Riemann surface.

In this section another important property of the ramification ideal comes to light: its norm $N(\mathfrak{z})$ is the discriminant D of the field Ω. In this way D acquires added meaning as the norm of an ideal.

If one recalls the classical concept of discriminant, as a function of polynomials of degree n that vanishes precisely for polynomials with less than n distinct roots, then one can expect a connection between the discriminant and ramification. Assuming that Ω somehow gives rise to a "Riemann surface" of n sheets, the discriminant ought to vanish where the n sheets are not distinct, that is, at ramification points.

Continuing this train of thought, we might expect the *degree* of the discriminant to give the *number* of ramification points. This is shown to be the case in §16.

1. *Lemma.* If any two of the ideals $\mathfrak{a}, \mathfrak{b}, \mathfrak{c}, \ldots$ are relatively prime, then there is a function which, modulo each of them, is congruent to a given function in $\mathfrak{o}$.[33]
Proof. One sets

$$\mathfrak{m} = \mathfrak{a}\mathfrak{b}\mathfrak{c}\cdots = \mathfrak{a}\mathfrak{a}_1 = \mathfrak{b}\mathfrak{b}_1 = \mathfrak{c}\mathfrak{c}_1 = \cdots.$$

---

[33]A form of the result generally known as the Chinese Remainder Theorem. (Translator's note.)

The greatest common divisor of $\mathfrak{a}_1 = \mathfrak{bc} \cdots, \mathfrak{b}_1 = \mathfrak{ac} \cdots, \mathfrak{c}_1 = \mathfrak{ab} \cdots$ is then equal to $\mathfrak{o}$, since no prime ideal can simultaneously divide $\mathfrak{a}_1, \mathfrak{b}_1, \mathfrak{c}_1, \ldots$. Hence (§4, 5) one can choose α_1 from $\mathfrak{a}_1$, β_1 from $\mathfrak{b}_1$, γ_1 from $\mathfrak{c}_1$, $\ldots$ so that

$$\alpha_1 + \beta_1 + \gamma_1 + \cdots = 1,$$

hence

$$
\begin{aligned}
\alpha_1 \equiv 1, \quad \beta_1 \equiv 0, \quad \gamma_1 \equiv 0, \quad \ldots \quad (\mathrm{mod}\ \mathfrak{a}), \\
\alpha_1 \equiv 0, \quad \beta_1 \equiv 1, \quad \gamma_1 \equiv 0, \quad \ldots \quad (\mathrm{mod}\ \mathfrak{b}), \\
\alpha_1 \equiv 0, \quad \beta_1 \equiv 0, \quad \gamma_1 \equiv 1, \quad \ldots \quad (\mathrm{mod}\ \mathfrak{c}), \\
\end{aligned}
$$

$$\cdots\cdots\cdots\cdots\cdots\cdots\cdots\cdots\cdots\cdots\cdots\cdots\cdots$$

Therefore, if $\lambda, \mu, \nu, \ldots$ are given functions in $\mathfrak{o}$,

$$\omega \equiv \lambda \alpha_1 + \mu \beta_1 + \nu \gamma_1 + \cdots \quad (\mathrm{mod}\ \mathfrak{m})$$

satisfies the conditions

$$\omega \equiv \lambda \ (\mathrm{mod}\ \mathfrak{a}), \quad \omega \equiv \mu \ (\mathrm{mod}\ \mathfrak{b}), \quad \omega \equiv \nu \ (\mathrm{mod}\ \mathfrak{c}), \quad \ldots \quad .$$

2. Let $\mathfrak{p}, \mathfrak{p}_1, \mathfrak{p}_2, \ldots$ be the distinct prime ideals dividing an arbitrary linear function $z - c$, and let

$$\mathfrak{o}(z - c) = \mathfrak{p}^e \mathfrak{p}_1^{e_1} \mathfrak{p}_2^{e_2} \cdots, \quad e + e_1 + e_2 + \cdots = n \quad (\S 9, 7).$$

One chooses functions $\lambda, \lambda_1, \lambda_2, \ldots$ divisible respectively by $\mathfrak{p}, \mathfrak{p}_1, \mathfrak{p}_2, \ldots$ but not by $\mathfrak{p}^2, \mathfrak{p}_1^2, \mathfrak{p}_2^2, \ldots$ and lets $b, b_1, b_2, \ldots$ be arbitrary but distinct constants. Then by 1 it is possible to determine a function ζ satisfying the congruences

$$\zeta \equiv b + \lambda \ (\mathrm{mod}\ \mathfrak{p}^2), \quad \zeta \equiv b_1 + \lambda_1 \ (\mathrm{mod}\ \mathfrak{p}_1^2), \quad \zeta \equiv b_2 + \lambda_2 \ (\mathrm{mod}\ \mathfrak{p}_2^2), \quad \ldots \quad ,$$

hence

$$\zeta \equiv b \ (\mathrm{mod}\ \mathfrak{p}), \quad \zeta \equiv b_1 \ (\mathrm{mod}\ \mathfrak{p}_1), \quad \zeta \equiv b_2 \ (\mathrm{mod}\ \mathfrak{p}_2), \quad \ldots \quad ,$$

so that, if a denotes any one of the constants, $\zeta - a$ will be divisible by at most one of the prime ideals $\mathfrak{p}, \mathfrak{p}_1, \mathfrak{p}_2, \ldots$, and never by its square. Therefore, if $\varphi(t) = \prod(t - a)$ is an integral algebraic function of the variable t with constant coefficients, then $\varphi(\zeta) = \prod(\zeta - a)$ is divisible by $\mathfrak{p}^m$ if and only if $\varphi(t)$ is algebraically divisible by $(t - b)^m$, and when $\mathfrak{p}^m$ is the highest power of $\mathfrak{p}$ dividing $\varphi(\zeta)$, $\mathfrak{p}^{m-1}$ is the highest power of $\mathfrak{p}$ dividing $\varphi'(\zeta)$. Consequently, $\varphi(\zeta)$ is divisible by $z - c$ only if $\varphi(t)$ is divisible by the nth degree function

$$\psi(t) = (t - b)^e (t - b_1)^{e_1} (t - b_2)^{e_2} \cdots .$$

It follows that the congruence

$$x_0 + x_1 \zeta + x_2 \zeta^2 + \cdots + x_{n-1} \zeta^{n-1} \equiv 0 \ (\mathrm{mod}\ z - c)$$

is satisfied only by those polynomial functions x that are all divisible by $z - c$. Thus if one sets

$$
\begin{aligned}
1 &= k_1^{(0)} \omega_1 + k_2^{(0)} \omega_2 + \cdots + k_n^{(0)} \omega_n, \\
\zeta &= k_1^{(1)} \omega_1 + k_2^{(1)} \omega_2 + \cdots + k_n^{(1)} \omega_n, \\
\zeta^2 &= k_1^{(2)} \omega_1 + k_2^{(2)} \omega_2 + \cdots + k_n^{(2)} \omega_n, \\
\end{aligned}
$$

$$\cdots\cdots\cdots\cdots\cdots\cdots\cdots\cdots\cdots\cdots\cdots$$

$$\zeta^{n-1} = k_1^{(n-1)} \omega_1 + k_2^{(n-1)} \omega_2 + \cdots + k_n^{(n-1)} \omega_n,$$

where $k_1^{(0)}, k_1^{(1)}, \ldots$ are polynomial functions of z and $\omega_1, \omega_2, \ldots, \omega_n$ is a basis of $\mathfrak{o}$, then the determinant

$$k = \sum \pm k_1^{(0)} k_2^{(1)} \cdots k_n^{(n-1)}$$

either vanishes identically or is divisible by $z - c$ (cf. the note to §9, 1).

Thus it follows that

$$N(t - \zeta) = f(t, z)$$

is irreducible. Now since $f(\zeta, z) = 0$, so that $f(\zeta, c)$ is divisible by $z - c$, $f(t, c)$ must be divisible by $\psi(t)$. Then, since both functions are of the same degree,

$$f(t, c) = \psi(t),$$

whence one concludes (for later use) that

$$Tr(\zeta) \equiv eb + e_1 b_1 + e_2 b_2 + \cdots \pmod{z - c}$$

and, applying the same considerations to the functions $\zeta^2, \zeta^3, \ldots$, which is certainly valid if the constants b are all nonzero,

$$Tr(\zeta^2) \equiv eb^2 + e_1 b_1^2 + e_2 b_2^2 + \cdots \pmod{z - c},$$
$$Tr(\zeta^3) \equiv eb^3 + e_1 b_1^3 + e_2 b_2^3 + \cdots \pmod{z - c},$$

$$\dotfill$$

Thus if $\mathfrak{p}^e$ is the highest power of $\mathfrak{p}$ dividing $f(\zeta, c)$, in which case $\mathfrak{p}^{e-1}$ is the highest power dividing $f'(\zeta, c)$, one has, since

$$f'(\zeta, c) \equiv f'(\zeta, z) \pmod{\mathfrak{p}^e},$$

that $\mathfrak{p}^{e-1}$ is also the highest power of $\mathfrak{p}$ dividing $f'(\zeta, z)$. This implies that

$$\mathfrak{o} f'(\zeta, z) = \mathfrak{m} \mathfrak{p}^{e-1} \mathfrak{p}_1^{e_1 - 1} \cdots,$$

where $\mathfrak{m}$, and hence $N(\mathfrak{m})$, is relatively prime to $z - c$.

Now if D is the discriminant of Ω it follows from this, from §10, (11), and from §2, (13) that (apart from constant factors)

$$N f'(\zeta, z) = \Delta(1, \zeta, \zeta^2, \ldots, \zeta^{n-1}) = D k^2 = (z - c)^{n-s} N(\mathfrak{m}),$$

where s is the number of distinct prime ideals $\mathfrak{p}, \mathfrak{p}_1, \mathfrak{p}_2, \ldots$ dividing $z - c$. And since k and $N(\mathfrak{m})$ are not divisible by $z - c$, $(z - c)^{n-s}$ is the highest power of $z - c$ dividing D. Consequently:

$$(1) \qquad\qquad D = \prod (z - c)^{n-s},$$

where the product is taken over all linear expressions $z - c$ divisible by less than n *distinct* prime factors, and hence divisible by a second or higher power of a prime ideal.

Thus there are only a finite number of linear functions $z - c$ divisible by the square of a prime ideal.

We now set

$$(2) \qquad\qquad \mathfrak{z} = \prod \mathfrak{p}^{e-1},$$

where the product is taken over all the prime ideals $\mathfrak{p}$ divisible by a power of their norms higher than the first—namely, the eth—and call this ideal $\mathfrak{z}$ the *ramification ideal*. It follows immediately from (1) and (2) that

$$(3) \qquad\qquad N(\mathfrak{z}) = D.$$

Since we also have $n - s \geq e - 1$, so $e(n - s) - 2(e - 1) \geq (e - 1)(e - 2) \geq 0$, D is divisible by $\mathfrak{p}^{2(e-1)}$, hence also by $\mathfrak{z}^2$, so we can set

$$\text{(4)} \qquad\qquad \mathfrak{o}D = \mathfrak{d}\mathfrak{z}^2, \quad N(\mathfrak{d}) = D^{n-2},$$

where $\mathfrak{d}$ also denotes an ideal.

3. If a function ρ in $\mathfrak{o}$ is divisible by each prime ideal dividing $z - c$, then $Tr(\rho)$ is divisible by $z - c$.

Proof. Let ζ be the same function as in 2, so that one can set:

$$x\rho = x_0 + x_1\zeta + x_2\zeta^2 + \cdots + x_{n-1}\zeta^{n-1},$$

where the coefficients $x, x_0, x_1, \ldots, x_{n-1}$ are polynomial functions of z without common divisor, the first of which is not divisible by $z - c$ (cf. 2). It follows from our hypothesis on the function ρ, and taking the constants b with the same meaning as in 2, that

$$x_0 + x_1 b + x_2 b^2 + \cdots + x_{n-1} b^{n-1} \equiv 0 \pmod{z - c},$$
$$x_0 + x_1 b_1 + x_2 b_2^2 + \cdots + x_{n-1} b_1^{n-1} \equiv 0 \pmod{z - c},$$
$$x_0 + x_1 b_2 + x_2 b_2^2 + \cdots + x_{n-1} b_2^{n-1} \equiv 0 \pmod{z - c},$$
$$\cdots\cdots\cdots\cdots\cdots\cdots\cdots\cdots\cdots\cdots\cdots\cdots\cdots\cdots\cdots$$

and this implies, multiplying the congruences by $e, e_1, e_2, \ldots$ and adding,

$$x_0 n + x_1 Tr(\zeta) + x_2 Tr(\zeta^2) + \cdots + x_{n-1} Tr(\zeta^{n-1}) = xTr(\rho) \equiv 0 \pmod{z - c}.$$

Therefore, since x is not divisible by $z - c$,

$$Tr(\rho) \equiv 0 \pmod{z - c}.$$

$$\text{Q.E.D.}$$

4. Now let

$$r = (z - c)(z - c_1)(z - c_2) \cdots$$

be the product of all the distinct linear factors of D and let

$$\mathfrak{r} = \mathfrak{p}\mathfrak{p}_1\mathfrak{p}_2 \cdots$$

be the product of all the distinct prime ideals dividing r. If $\mathfrak{z}$ is the ramification ideal as above then

$$\text{(5)} \qquad\qquad \mathfrak{r}\mathfrak{z} = \prod \mathfrak{p}^e = \mathfrak{o}r$$

and hence

$$N(\mathfrak{r}) = \frac{r^n}{D}.$$

By 3, each function ρ in $\mathfrak{r}$ has the property that $Tr(\rho)$ is divisible by r. Now if, as in §10,

$$\mathfrak{e} = [\varepsilon_1, \varepsilon_2, \ldots, \varepsilon_n]$$

is the module complementary to $\mathfrak{o}$, and if ρ is an arbitrary function in $\mathfrak{r}$, then one can set

$$\rho = x_1\varepsilon_1 + x_2\varepsilon_2 + \cdots + x_n\varepsilon_n,$$

where

$$x_\iota = Tr(\rho\omega_\iota),$$

by §10, 3. Therefore, since $\rho\omega_\iota$ is a function in $\mathfrak{r}$, x_ι is a polynomial function of z divisible by r. It follows that the ideal $\mathfrak{r}$ is divisible by the module $r\mathfrak{e}$. Hence the ideal $D\mathfrak{r}$ is also divisible by the ideal $rD\mathfrak{e}$. At the same time,

$$N(D\mathfrak{r}) = r^n D^{n-1}, \quad N(rD\mathfrak{e}) = r^n D^{n-1} \quad (\S 10, 10),$$

hence by §8, 9,

$$D\mathfrak{r} = rD\mathfrak{e}$$

or

(6) $$\mathfrak{r} = r\mathfrak{e}.$$

It follows from this and the above remark about ρ that, for any function ε in $\mathfrak{e}$, $Tr(\varepsilon)$ is a *polynomial* function of z. Multiplying (6) by $\mathfrak{z}$ and applying (5) we get

$$\mathfrak{r}\mathfrak{z} = r\mathfrak{e}\mathfrak{z} = \mathfrak{o}r$$

and hence

(7) $$\mathfrak{e}\mathfrak{z} = \mathfrak{o}.$$

If one multiplies the latter equation by D, then it follows from (4) that

$$\mathfrak{e}D\mathfrak{z} = \mathfrak{z}^2\mathfrak{d},$$

hence

(8) $$D\mathfrak{e} = \mathfrak{z}\mathfrak{d},$$

and multiplying the latter equation by $\mathfrak{e}$ and using (7),

(9) $$D\mathfrak{e}^2 = \mathfrak{d}.$$

5.[34] If θ is an integral algebraic function of z in Ω and $N(t - \theta) = f(t)$, then $f'(\theta)$ is divisible by the ramification ideal $\mathfrak{z}$.

Proof. If $f(t)$ is reducible then $f'(\theta) = 0$, hence it is certainly divisible by $\mathfrak{z}$. Otherwise, by §10, 11

$$\mathfrak{e}f'(\theta) = \mathfrak{f}$$

is an ideal, hence, multiplying by $\mathfrak{z}$ and using (7),

(10) $$\mathfrak{o}f'(\theta) = \mathfrak{f}\mathfrak{z}.$$

At the same time, it follows by substituting

$$\theta^r = \sum_\iota k_\iota^{(r)}\omega_\iota,$$

$$k = \sum \pm k_1^{(0)} k_2^{(1)} \cdots k_{(n)}^{(n-1)}$$

in §10, 11 that

$$N f'(\theta) = N(\mathfrak{f})N(\mathfrak{z}) = DN(\mathfrak{f})$$
$$= \text{const.}k^2 D \qquad (\S 10, (11) \text{ and } \S 2, (13)),$$

so

(11) $$N(\mathfrak{f}) = \text{const.}k^2$$

is a perfect square.

---

[34]This subsection is wrongly numbered 4 in the original. However, Dedekind and Weber have correctly made cross references to §11, 5. (Translator's note.)

§12. The fractional functions of z in the field Ω

SUMMARY AND COMMENTS

Dedekind and Weber now extend their considerations from "integers" of the function field Ω to arbitrary members. These may be viewed as quotients of "integers," and indeed the denominator may be taken to be a polynomial. (This is analogous to the situation in algebraic number fields, where any element of the field is the quotient of an algebraic integer by an ordinary integer. See Dedekind (1877), §18.)

Quotients of ideals, defined for modules in general in §4, serve to represent any principal ideal $\mathfrak{o}\eta$. If $\mathfrak{o}\eta = \frac{\mathfrak{b}}{\mathfrak{a}}$ then $\mathfrak{b}$ is called the *upper* ideal, and $\mathfrak{a}$ the *lower* ideal, of the function η. The upper and lower ideals encode the zeros and poles, respectively, of η.

If ρ is a function such that $\mathfrak{o}\rho = \frac{\mathfrak{m}\mathfrak{p}}{\mathfrak{n}}$, where $\mathfrak{m}$ and $\mathfrak{n}$ are not divisible by the prime ideal $\mathfrak{p}$, and if η is a function whose lower ideal is not divisible by $\mathfrak{p}$, then η may be expressed as a polynomial in ρ. Weber (1908), §185, in his version of the Dedekind-Weber theory, called this the "Taylor expansion" of η.

1. Each function η in Ω can be expressed as a quotient of two integral algebraic functions of z (§3, 3), in infinitely many ways (the denominator can even be a polynomial function of z). Suppose therefore that

$$\eta = \frac{\nu}{\mu},$$

where μ, ν are integral algebraic functions of z (functions in $\mathfrak{o}$). Now if $\mathfrak{m}$ is the greatest common divisor of the two principal ideals $\mathfrak{o}\mu, \mathfrak{o}\nu$ then

(1) $$\mathfrak{o}\mu = \mathfrak{a}\mathfrak{m}, \quad \mathfrak{o}\nu = \mathfrak{b}\mathfrak{m},$$

where $\mathfrak{a}, \mathfrak{b}$ are relatively prime ideals, and hence by §4, 6,

(2) $$\mathfrak{a}\nu = \mathfrak{b}\mu \quad \text{or} \quad \mathfrak{a}\eta = \mathfrak{b}.$$

Thus if α is any function in $\mathfrak{a}$, then $\alpha\eta$ is in $\mathfrak{b}$, hence it is also an integral algebraic function of z. Conversely, if α is an integral algebraic function of z with the property that $\alpha\eta = \beta$ is an integral algebraic function, then

$$\alpha\nu = \beta\mu$$

and hence by (1)

$$\alpha\mathfrak{b} = \beta\mathfrak{a}.$$

Now since $\mathfrak{a}, \mathfrak{b}$ are relatively prime, α must be divisible by $\mathfrak{a}$, and β by $\mathfrak{b}$, which implies:

$\mathfrak{a}$ is the collection of all those integral algebraic functions α with the property that $\alpha\eta$ is an integral algebraic function, and the collection of all these integral algebraic functions $\alpha\eta$ is the ideal $\mathfrak{b}$; in other words:

$\mathfrak{b}$ *is the least common multiple of* $\mathfrak{o}\eta$ *and* $\mathfrak{o}$, *and* $\mathfrak{a}$ *is the least common multiple of* $\frac{\mathfrak{o}}{\eta}$ *and* $\mathfrak{o}$. It follows that if $\mathfrak{a}', \mathfrak{b}'$ are two ideals satisfying the condition

$$\mathfrak{a}'\eta = \mathfrak{b}',$$

then $\mathfrak{a}'$ must be divisible by $\mathfrak{a}$. Suppose therefore that

$$\mathfrak{a}' = \mathfrak{n}\mathfrak{a};$$

then

$$\mathfrak{b}' = \mathfrak{n}\mathfrak{a}\eta = \mathfrak{n}\mathfrak{b}.$$

Conversely, for any ideal $\mathfrak{n}$,

$$\mathfrak{n}\mathfrak{a}\eta = \mathfrak{n}\mathfrak{b}.$$

2. Now let $\mathfrak{a}, \mathfrak{b}$ be two ideals satisfying the condition

$$\mathfrak{a}\eta = \mathfrak{b},$$

whether they are relatively prime or not. By §4, 8 the quotient $\frac{\mathfrak{b}}{\mathfrak{a}}$ is the collection of all those functions γ with the property that $\mathfrak{a}\gamma$ is divisible by $\mathfrak{b}$. These functions certainly include all functions of the form $\omega\eta$, where ω is any function in $\mathfrak{o}$. But conversely, each function γ is also of this form. Since $\mathfrak{a}\gamma$ is divisible by $\mathfrak{b}$, and hence also by $\mathfrak{o}$, it is an ideal (because it has the properties I, II of §7). Thus

$$\mathfrak{a}\gamma = \mathfrak{c}\mathfrak{b},$$

where $\mathfrak{c}$ is likewise an ideal and, multiplying by η,

$$\mathfrak{b}\gamma = \mathfrak{c}\mathfrak{b}\eta.$$

Now if $\eta = \frac{\nu}{\mu}$ as above, and if ρ, σ are integral algebraic functions with $\gamma = \frac{\rho}{\sigma}$, then it follows that

$$\mathfrak{b}\rho\mu = \mathfrak{c}\mathfrak{b}\nu\sigma,$$

hence

$$\mathfrak{o}\rho\mu = \mathfrak{c}\nu\sigma, \quad \mathfrak{o}\gamma = \mathfrak{c}\eta.$$

The two together yield the theorem

$$(3) \qquad\qquad \mathfrak{o}\eta = \frac{\mathfrak{b}}{\mathfrak{a}}.$$

If $\mathfrak{b}, \mathfrak{a}$ are relatively prime in this representation, which by 1 is always possible in precisely one way, then $\mathfrak{b}$ will be called the *upper ideal*, and $\mathfrak{a}$ the *lower ideal*, of the function η.

3. Again, if

$$\mathfrak{a}\eta = \mathfrak{b}, \quad \text{so that} \quad \mathfrak{o}\eta = \frac{\mathfrak{b}}{\mathfrak{a}},$$

and if α is any function in $\mathfrak{a}$ and β any function in $\mathfrak{b}$, then

$$\eta = \frac{\beta}{\alpha} \quad \text{and} \quad \mathfrak{a}\beta = \mathfrak{b}\alpha.$$

It follows by taking norms that

$$N(\eta) = \text{const.}\frac{N(\mathfrak{a})}{N(\mathfrak{b})}.$$

4. If η, η' are two functions in ω and if

$$\mathfrak{a}\eta = \mathfrak{b}, \quad \mathfrak{a}'\eta' = \mathfrak{b}'$$

as in 1, but regardless of whether $\mathfrak{a}, \mathfrak{b}$ or $\mathfrak{a}', \mathfrak{b}'$ are relatively prime, then

$$\mathfrak{a}\mathfrak{a}'\eta\eta' = \mathfrak{b}\mathfrak{b}'.$$

The equations

$$\mathfrak{o}\eta = \frac{\mathfrak{b}}{\mathfrak{a}}, \quad \mathfrak{o}\eta' = \frac{\mathfrak{b}'}{\mathfrak{a}'}$$

therefore yield the equations

$$\mathfrak{o}\eta\eta' = \frac{\mathfrak{b}\mathfrak{b}'}{\mathfrak{a}\mathfrak{a}'}, \quad \mathfrak{o}\frac{1}{\eta} = \frac{\mathfrak{a}}{\mathfrak{b}}, \quad \mathfrak{o}\frac{\eta}{\eta'} = \frac{\mathfrak{b}\mathfrak{a}'}{\mathfrak{a}\mathfrak{b}'}.$$

5. If $\mathfrak{a}\eta = \mathfrak{b}, \mathfrak{a}\eta' = \mathfrak{b}'$ then also

$$\mathfrak{a}(\eta \pm \eta') = \mathfrak{b}''$$

will be an ideal, because when $\alpha\eta, \alpha\eta'$ are integral algebraic functions so too is $\alpha(\eta \pm \eta')$. Thus if

$$\mathfrak{o}\eta = \frac{\mathfrak{b}}{\mathfrak{a}}, \quad \mathfrak{o}\eta' = \frac{\mathfrak{b}'}{\mathfrak{a}},$$

then it follows that

$$\mathfrak{o}(\eta \pm \eta') = \frac{\mathfrak{b}''}{\mathfrak{a}}.$$

If the two ideals $\mathfrak{b}, \mathfrak{b}'$ have a common divisor, then the latter is also a divisor of $\mathfrak{b}''$.

6. Now let ρ be a function in Ω whose upper ideal is divisible by a prime ideal $\mathfrak{p}$, but not by $\mathfrak{p}^2$. (Such functions always exist. They can even be integral algebraic functions of z.) Thus

$$\mathfrak{o}\rho = \frac{\mathfrak{m}\mathfrak{p}}{\mathfrak{n}},$$

where $\mathfrak{m}, \mathfrak{n}$ are ideals not divisible by $\mathfrak{p}$. Suppose also that η is any function in Ω whose lower ideal is not divisible by $\mathfrak{p}$, so that

$$\mathfrak{o}\eta = \frac{\mathfrak{b}}{\mathfrak{a}}$$

where $\mathfrak{a}$ is not divisible by $\mathfrak{p}$. One chooses an arbitrary function α in $\mathfrak{a}$ that is not divisible by $\mathfrak{p}$, and a corresponding function β in $\mathfrak{b}$ so that

$$\eta = \frac{\beta}{\alpha}.$$

Let

$$\alpha \equiv \alpha_0, \quad \beta \equiv \beta_0 \pmod{\mathfrak{p}}, \quad c_0 = \frac{\beta_0}{\alpha_0},$$

where α_0, β_0, c_0 are constants, the first of which is nonzero. By 5 we have

$$\mathfrak{o}(\eta - c_0) = \mathfrak{o}\frac{\beta - c_0\alpha}{\alpha} = \frac{\mathfrak{b}_1}{\mathfrak{a}},$$

and it follows from

$$\mathfrak{a}(\beta - c_0\alpha) = \mathfrak{b}_1\mathfrak{a}, \quad \beta - c_0\alpha \equiv 0 \pmod{\mathfrak{p}},$$

and the fact that α is not divisible by $\mathfrak{p}$, that $\mathfrak{b}_1$ must be divisible by $\mathfrak{p}$. Thus if one sets

$$\eta - c_0 = \rho\eta_1,$$

then the lower ideal of η_1 is also not divisible by $\mathfrak{p}$. In this way a whole series of constants $c_0, c_1, \ldots, c_{r-1}, \ldots$ may be determined so that

$$\eta = c_0 + \rho\eta_1,$$
$$\eta_1 = c_1 + \rho\eta_2,$$
$$\cdots\cdots\cdots$$
$$\eta_{r-1} = c_{r-1} + \rho\eta_r, \quad \cdots$$

where $\eta_1, \eta_2, \ldots, \eta_r, \ldots$ are functions whose lower ideals have no prime factors other than those of the lower ideal of η and the upper ideal of ρ, with $\mathfrak{p}$ excluded. Accordingly, for each positive integer r,

$$\eta = c_0 + c_1\rho + \cdots + c_{r-1}\rho^{r-1} + \eta_r\rho^r.$$

If the lower ideal of ζ is divisible by $\mathfrak{p}^s$, but not by $\mathfrak{p}^{s+1}$, then one can apply these considerations to the function $\eta = \zeta\rho^s$ and obtain

$$\zeta = c_0\rho^{-s} + c_1\rho^{-s+1} + \cdots + c_{r-1}\rho^{-s+r} + \eta_r\rho^{-s+r}.$$

§13. Rational transformations of functions in the field Ω

SUMMARY AND COMMENTS

It was shown in §2 that any nonconstant function $z_1 \in \Omega$ satisfies a polynomial equation whose coefficients are rational functions of z. It follows that there is a polynomial equation relating z and z_1. This suggests that z_1 may serve as an "independent variable" for the functions in Ω just as well as z, and indeed this is the case.

If θ is a function such that $1, \theta, \theta^2, \ldots, \theta^{n-1}$ is a basis of Ω (relative to z), then we can find a function θ_1 such that $1, \theta_1, \theta_1^2, \ldots, \theta_1^{n-1}$ is a basis of Ω relative to z_1. Moreover, z_1, θ_1 can be expressed rationally in terms of z, θ, and vice versa. This is the "rational transformation" referred to in the title of the section, though it might be better called (as it is today) a *birational* transformation.

The section concludes with the warning that "the concepts of basis, norm, trace, $\ldots$, ideal depend essentially on the choice of independent variable," except when "the two variables z, z_1 depend linearly on each other." The latter sense of "linear" is that $z_1 = \frac{az+b}{cz+d}$ for some numbers a, b, c, d with $ad - bc \neq 0$.

If z_1 is an arbitrary nonconstant function in the field Ω (a *variable* in Ω) then, as we have shown in §2, there is an irreducible algebraic equation relating z_1 and z, which, when cleared of denominators, is of degree e in z_1 and degree e_1 in z. As was also shown there, e is a divisor of n, $n = ef$. Let this equation be

(1) $$G(\overset{e}{z_1}\overset{e_1}{z}) = 0.$$

With the help of this equation, each rational function ζ of z and z_1 may (§1) be brought into the two forms

(2) $$\begin{cases} \zeta = x_0 + x_1 z_1 + \cdots + x_{e-1} z_1^{e-1}, \\ \zeta = x_0^{(1)} + x_1^{(1)} z + \cdots + x_{e_1-1}^{(1)} z^{e_1-1}, \end{cases}$$

and indeed uniquely, where $x_0, x_1, \ldots, x_{e-1}$ are rational functions of z and $x_0^{(1)}, x_1^{(1)},$ $\ldots, x_{e_1-1}^{(1)}$ are rational functions of z_1.

Now if θ is a function such that $1, \theta, \theta^2, \ldots, \theta^{n-1}$ is a basis[35] of Ω (relative to z), then by §2 the n functions

(3) $\quad \begin{cases} 1, & z_1, & z_1^2, & \ldots, & z_1^{e-1}, \\ \theta, & \theta z_1, & \theta z_1^2, & \ldots, & \theta z_1^{e-1}, \\ \ldots & \ldots & \ldots & \ldots & \ldots \\ \theta^{f-1}, & \theta^{f-1} z_1, & \theta^{f-1} z_1^2, & \ldots, & \theta^{f-1} z_1^{e-1} \end{cases}$

likewise form such a basis, and hence it follows from (2) that the $e_1 f = n_1$ functions

(4) $\quad \begin{cases} 1, & z, & z^2, & \ldots, & z^{e_1-1}, \\ \theta, & \theta z, & \theta z^2, & \ldots, & \theta z^{e_1-1}, \\ \ldots & \ldots & \ldots & \ldots & \ldots \\ \theta^{f-1}, & \theta^{f-1} z, & \theta^{f-1} z^2, & \ldots, & \theta^{f-1} z^{e_1-1}, \end{cases}$

which we abbreviate by

$$\eta_1^{(1)}, \quad \eta_2^{(1)}, \quad \ldots, \quad \eta_{n_1}^{(1)},$$

satisfy an equation of the form

$$x_1^{(1)} \eta_1^{(1)} + x_2^{(1)} \eta_2^{(1)} + \cdots + x_{n_1}^{(1)} \eta_{n_1}^{(1)} = 0$$

only if the rational functions $x_1^{(1)}, x_2^{(1)}, \ldots, x_{n_1}^{(1)}$ of z_1 all vanish. It then follows by (2) again that each function η in Ω is representable uniquely in the form

$$\eta = x_1^{(1)} \eta_1^{(1)} + x_2^{(1)} \eta_2^{(1)} + \cdots + x_{n_1}^{(1)} \eta_{n_1}^{(1)},$$

where the $x^{(i)}$ are rational functions of z_1.

Each such function η satisfies an algebraic equation of degree n_1 whose coefficients depend rationally on z_1, because

$$\eta \eta_1^{(1)} = x_{1,1}^{(1)} \eta_1^{(1)} + x_{1,2}^{(1)} \eta_2^{(1)} + \cdots + x_{1,n_1}^{(1)} \eta_{n_1}^{(1)},$$
$$\eta \eta_2^{(1)} = x_{2,1}^{(1)} \eta_1^{(1)} + x_{2,2}^{(1)} \eta_2^{(1)} + \cdots + x_{2,n_1}^{(1)} \eta_{n_1}^{(1)},$$
$$\cdots\cdots\cdots\cdots\cdots\cdots\cdots\cdots\cdots\cdots\cdots$$
$$\eta \eta_{n_1}^{(1)} = x_{n_1,1}^{(1)} \eta_1^{(1)} + x_{n_1,2}^{(1)} \eta_2^{(1)} + \cdots + x_{n_1,n_1}^{(1)} \eta_{n_1}^{(1)},$$

and hence

$$\begin{vmatrix} x_{1,1}^{(1)} - \eta & x_{1,2}^{(1)} & \cdots & x_{1,n_1}^{(1)} \\ x_{2,1}^{(1)} & x_{2,2}^{(1)} - \eta & \cdots & x_{2,n_1}^{(1)} \\ \cdots & \cdots & \cdots & \cdots \\ x_{n_1,1}^{(1)} & x_{n_1,2}^{(1)} & & x_{n_1,n_1}^{(1)} - \eta \end{vmatrix} = 0.$$

It may now be shown that the function $\eta = \theta_1$ can be chosen so that θ_1 satisfies no equation of lower degree whose coefficients depend rationally on z_1.

To prove this assertion we rely on the following theorem, which is easily proved by induction on m. If

$$F(x_1, x_2, \ldots, x_m)$$

is a polynomial function of $x_1, x_2, \ldots, x_m$ whose coefficients are functions in Ω, not all vanishing, then one can replace $x_1, x_2, \ldots, x_m$ by constants or rational functions

<hr />

[35]One could take any other basis of Ω in place of $1, \theta, \ldots, \theta^{n-1}$ for this argument. However, it suits our purpose to choose this particular one.

of z_1 in such a way that F becomes a nonvanishing function in Ω. Because if $F(x_1, x_2, \ldots, x_m)$ is zero for *all* such $x_1, x_2, \ldots, x_m$ then also

$$dF = F'(x_1)dx_1 + F'(x_2)dx_2 + \cdots + F'(x_m)dx_m = 0$$

for $dx_1, dx_2, \ldots, dx_m$ which are arbitrary constants or rational functions of z_1. Now if

$$\theta_1 = x_1 \eta_1^{(1)} + x_2 \eta_2^{(1)} + \cdots + x_{n_1} \eta_{n_1}^{(1)}$$

and

(5)
$$\begin{cases} 1 = x_{1,0}\eta_1^{(1)} + x_{2,0}\eta_2^{(1)} + \cdots + x_{n_1,0}\eta_{n_1}^{(1)}, \\ \theta_1 = x_{1,1}\eta_1^{(1)} + x_{2,1}\eta_2^{(1)} + \cdots + x_{n_1,1}\eta_{n_1}^{(1)}, \\ \cdots\cdots\cdots\cdots\cdots\cdots\cdots\cdots\cdots\cdots\cdots\cdots\cdots \\ \theta_1^m = x_{1,m}\eta_1^{(1)} + x_{2,m}\eta_n^{(1)} + \cdots + x_{n_1,m}\eta_{n_1}^{(1)}, \end{cases}$$

then the $x_{k,h}$ are homogeneous polynomials of degree h in $x_1, x_2, \ldots, x_{n_1}$ and hence they depend rationally on z_1.

Thus if

$$\varphi(\theta_1) = a_m \theta_1^m + a_{m-1}\theta_1^{m-1} + \cdots + a_1 \theta_1 + a_0 = 0$$

is the equation of lowest degree satisfied by θ_1 with coefficients depending rationally on z_1, then the functions $a_0, a_1, \ldots, a_m$ satisfy the conditions

$$a_0 x_{\iota,0} + a_1 x_{\iota,1} + \cdots + a_m x_{\iota,m} = 0 \qquad (\iota = 1, 2, \ldots, n_1)$$

and $m \leq n_1$. Now since not all the $m \times m$ determinants formed from the coefficients $x_{h,k}$ can vanish (otherwise θ_1 would satisfy an equation of degree less than m) it follows from the latter equations that one can assume the $a_0, a_1, \ldots, a_m$ to be homogeneous polynomial functions of $x_1, x_2, \ldots, x_{n_1}$.

Now if the equation $\varphi(\theta_1) = 0$ is to hold for all $x_1, x_2, \ldots, x_{n_1}$ depending rationally on z_1, the above theorem shows that we must also have

$$d\varphi = \varphi'(\theta_1)d\theta_1 + da_m\theta_1^m + \cdots + da_1\theta_1 + da_0 = 0,$$

and when $m < n_1$ the $dx_1, dx_2, \ldots, dx_{n_1}$ may be determined, without all vanishing, so that

$$da_m : da_{m-1} : \cdots : da_1 : da_0 = a_m : a_{m-1} : \cdots : a_1 : a_0$$

and hence

$$\varphi'(\theta_1)d\theta_1 = 0.$$

But since $\varphi'(\theta_1)$ is of degree $m-1$ we must have $d\theta_1 = 0$, hence $dx_1 = 0$, $dx_2 = 0$, $\ldots$, $dx_{n_1} = 0$. Therefore we can only have $m = n_1$.

Thus, if θ_1 is determined so that the equation of lowest degree

$$F_1(\overset{n_1}{\theta}_2, z_1) = 0$$

really attains the degree n_1, then all functions in Ω are uniquely representable in the form

$$\eta = x_0^{(1)} + x_1^{(1)}\theta_1 + \cdots + x_{n_1}^{(1)}\theta_1^{n_1-1},$$

where the coefficients $x_0^{(1)}, x_1^{(1)}, \ldots, x_{n_1-1}^{(1)}$ depend rationally on z_1. Because under this assumption one can represent $\eta_1^{(1)}, \eta_2^{(1)}, \ldots, \eta_{n_1}^{(1)}$ in this way by means of the equations (5).

Thus not only can z_1, θ_1 be expressed rationally in terms of z, θ but, conversely, z, θ can be expressed rationally in terms of z_1, θ_1.

The variable z that we previously took to be independent can therefore be any (nonconstant) function in the field Ω. However, while the collection of all functions in the field Ω remains completely unaltered, the concepts of *basis, norm, trace, discriminant, integral algebraic function, module* and *ideal* depend essentially on the choice of *independent* variable z.

Only in the special case where the two variables z, z_1 depend linearly on each other is a basis of Ω relative to z also a basis of Ω relative to z_1. Similarly, the norms, traces and discriminants are also the same for z and z_1 in this case.

If α, β are any two functions in Ω then they are related by polynomial equations in α and β.

Among these there is one,

$$F(\alpha, \beta) = 0,$$

that by §1 has lowest possible degree in both α and β, and this will be called the *irreducible equation relating α and β*. It is unique up to a constant factor.

Part II

§14. The points of the Riemann surface

SUMMARY AND COMMENTS

As Dedekind and Weber point out, up to this stage algebraic functions have been treated merely as elements of a field Ω, and their numerical values have not been considered. Of course, a function α acquires a value $\alpha(P)$ when we *evaluate α at the point P* (which we expect to be a point of a Riemann surface). But as yet we do not have any "points." Dedekind and Weber had the inspiration to *define points in terms of the function field Ω.*

Since functions acquire values from a point, a point $\mathfrak{P}$ corresponds to an *evaluation* of functions, that is, an assignment of values $(\alpha)_0$ to the functions $\alpha \in \Omega$ that is *consistent* in the following sense:

(I) $\alpha_0 = \alpha$ if α is constant, and in general
(II) $(\alpha + \beta)_0 = \alpha_0 + \beta_0,$ (IV) $(\alpha\beta)_0 = \alpha_0\beta_0,$
(III) $(\alpha - \beta)_0 = \alpha_0 - \beta_0,$ (V) $(\alpha/\beta)_0 = \alpha_0/\beta_0.$

Since we are dealing with a field, the value $1/0$ is going to come up, so we include ∞ among the possible values.

There is an immediate connection with prime ideals; namely, the integral algebraic functions that vanish at a point $\mathfrak{P}$ form a prime ideal $\mathfrak{p}$. Conversely, each prime ideal occurs as the set of functions vanishing at some point $\mathfrak{P}$.

These results depend on the choice of independent variable z but, with finitely many exceptions, the points $\mathfrak{P}$ are in one-to-one correspondence with the prime ideals $\mathfrak{p}$ of Ω. When finitely many other points $\mathfrak{P}'$ are added (corresponding to prime ideals $\mathfrak{p}'$ for the variable $z' = 1/z$) one has all the points of the *Riemann surface* for Ω.

One could say that the prime ideals of Ω correspond to all points of the Riemann surface except its "points at infinity." However, the points at infinity are important, so Dedekind and Weber in the next section introduce their concept of "polygons" (now called "divisors") in order to give a better accounting of points at infinity.

It should be mentioned that, at this stage, the Riemann surface for Ω is just a cloud of points without obvious structure. Some structure will be extracted algebraically later, but *not* the "continuity" or connectedness of this set of points. Because of the generality

of the Dedekind-Weber theory, which also applies to discrete function fields, this is unavoidable. But it means that the usual approach
to differential quotients and differentials, via limits, is not available.
Instead, Dedekind and Weber show in §23 that differentials can also
be defined algebraically.

The previous considerations of the functions in the field Ω were of a purely formal nature. All results were rational, i.e., consequences of the irreducible equation
relating the two functions in Ω by the four operations of algebra. The numerical
values of these functions did not come under consideration. One could even push
the formal treatment further, without invoking other principles, by regarding two
functions in the field Ω as independent variables instead of being related by an
equation, in which case everything boils down to algebraic divisibility of rational
functions of two variables. We have also worked out this approach, but it is very
heavy going and offers no advantage in rigor over the way we have taken above.
Now that we have carried the formal part of the investigation this far, the most
pressing question is: *to what extent is it possible to assign numerical values to the
functions in Ω so that all rational relations (identities) between these functions become correct equations between numbers?* For this investigation it turns out to be
useful to regard infinity as *one* definite (constant) number ∞, subject to definite
rules of calculation.[36] When the domain of numbers is extended in this way the
arithmetic operations always lead to a definite result as long as none of the terms
$\infty \pm \infty$, $0 \cdot \infty$, $\frac{0}{0}$, $\frac{\infty}{\infty}$ occurs in the course of the calculation. The appearance of
such indeterminacy in an equation is not to be regarded as a contradiction, since
in this case the equation no longer expresses a definite assertion, and hence cannot
be said to be true or untrue. The functions in the field Ω include not only infinitely
many variables but also all the constants, i.e., numbers. Thus in trying to satisfy
the above demand we arrive at the following concept.

1. *Definition.* When *all* individual elements $\alpha, \beta, \gamma, \ldots$ of the field Ω are replaced by *definite* numerical values $\alpha_0, \beta_0, \gamma_0, \ldots$ in such a way that

(I) $\alpha_0 = \alpha$ if α is constant, and in general
(II) $(\alpha + \beta)_0 = \alpha_0 + \beta_0$,   (IV) $(\alpha\beta)_0 = \alpha_0\beta_0$
(III) $(\alpha - \beta)_0 = \alpha_0 - \beta_0$,   (V) $(\alpha/\beta)_0 = \alpha_0/\beta_0$,

then such an assignment of definite values will be associated with a *point* $\mathfrak{P}$ (which
one may represent by any point in space),[37] and we say that $\alpha = \alpha_0$ at $\mathfrak{P}$, or that
α has the value α_0 *at* $\mathfrak{P}$. Two points are called different if and only if there is a
function α in Ω to which they assign different values.

From this definition of point we shall now deduce existence, as well as the
scope of the concept. However, it should first be emphasized that this definition of
"point" is an *invariant* concept for the field Ω, in no way depending on the choice
of independent variables used to represent the functions in the field.

---

[36]Regarding infinity as a definite value is common and useful in function theory. This was
done, e.g., by Riemann in order to be able to consider his surfaces representing algebraic functions
as closed.

[37]A geometric representation of the "point" is by no means necessary, and does not make
comprehension much easier. It suffices to regard the word "point" as a short and convenient
expression for the coexistence of values just described.

2. *Theorem.* If a point $\mathfrak{P}$ is given, and z is a variable in Ω that is finite at $\mathfrak{P}$ (such a variable exists for every point, because if $z_0 = \infty$ then $\left(\frac{1}{z}\right)_0 = 0$ is finite), then each integral algebraic function ω of z has a finite value ω_0 at $\mathfrak{P}$. This is because between ω and z there is a relation of the form

$$1 = a\frac{1}{\omega} + b\frac{1}{\omega^2} + \cdots + k\frac{1}{\omega^m},$$

where $a, b, \ldots, k$ are *polynomial* functions of z by (II), (III), (IV), and hence they have finite values at $\mathfrak{P}$. Consequently, $\left(\frac{1}{\omega_0}\right)$ is nonzero, and hence ω_0 cannot equal ∞.

3. *Theorem.* If z is a variable that is finite at $\mathfrak{P}$ then the set $\mathfrak{p}$ of all the integral algebraic functions π of z that vanish at $\mathfrak{P}$ is a prime ideal in z. We say that the point $\mathfrak{P}$ *generates* this prime ideal $\mathfrak{p}$. If ω is an integral algebraic function of z that has the value ω_0 at $\mathfrak{P}$ then $\omega \equiv \omega_0 \pmod{\mathfrak{p}}$.

Proof. If $\pi'_0, \pi''_0 = 0$ then also $(\pi' + \pi'')_0 = \pi'_0 + \pi''_0 = 0$, and if ω is any integral algebraic function of z, so that ω_0 is finite, then $\pi_0 = 0$ implies $(\omega\pi)_0 = \omega_0\pi_0 = 0$. Thus $\mathfrak{p}$ is an *ideal* in z (§7, I, II). The ideal $\mathfrak{p}$ is different from $\mathfrak{o}$, since it does not contain the function "1."

If ω has the value ω_0 at $\mathfrak{P}$, then $(\omega - \omega_0)_0 = 0$, hence $\omega \equiv \omega_0 \pmod{\mathfrak{p}}$, and so each integral algebraic function of z is congruent to a constant modulo $\mathfrak{p}$. Therefore (§9, 7) $\mathfrak{p}$ is a *prime ideal*.

4. *Theorem.* The same prime ideal $\mathfrak{p}$ cannot be generated by two different points.

Firstly, the value of each integral algebraic function ω at a point $\mathfrak{P}$ generating the ideal $\mathfrak{p}$ is determined by the congruence $\omega \equiv \omega_0 \pmod{\mathfrak{p}}$. But if η is an arbitrary function in Ω then it is possible, by §12, 1, to determine two integral algebraic functions α, β, not divisible by $\mathfrak{p}$, such that

$$\eta = \frac{\alpha}{\beta}.$$

Now, since the finite values α_0, β_0 do not both vanish, it follows from (V) that

$$\eta_0 = \frac{\alpha_0}{\beta_0}$$

is likewise determined by $\mathfrak{p}$.

It follows that two points at which a variable z has finite values are different if and only if there is an *integral algebraic* function of z taking different values at the points.

5. *Theorem.* If z is any variable in Ω and $\mathfrak{p}$ is a prime ideal in z then there is *one* (and, by 4, *only* one) point $\mathfrak{P}$ that generates this prime ideal, and which will be called the *null point* of the ideal $\mathfrak{p}$.

Proof. Let η be an arbitrary function in Ω, and let ρ be one whose upper ideal is divisible by $\mathfrak{p}$ but not by $\mathfrak{p}^2$. Then by §12, 6 there is a unique way to choose an integer m, a nonzero finite constant c, and a function η_1 whose lower ideal is not divisible by $\mathfrak{p}$, so that

$$\eta = c\rho^m + \eta_1\rho^{m+1}.$$

We set

$$\eta_0 = 0, \ c, \ \infty$$

according as m is positive, zero, or negative. This assignment of values to the functions in the field Ω corresponds to a point $\mathfrak{P}$, since conditions (I) to (V) are satisfied, as one sees immediately.[38]

Each function whose upper ideal is divisible by $\mathfrak{p}$, in particular, each function in $\mathfrak{p}$, receives the value zero at $\mathfrak{P}$ by this assignment, i.e., the point $\mathfrak{P}$ thus determined generates the prime ideal $\mathfrak{p}$.

The functions whose lower ideal is divisible by $\mathfrak{p}$, and only such functions, have the value ∞ at $\mathfrak{P}$. This means that an *integral algebraic* function of z is infinite at no point where z has a finite value and, since a fractional function of z certainly has one prime ideal in its lower ideal, and hence at least one point where z is finite and the function value is infinite, it follows conversely that each function with no infinite values for finite values of z is an integral algebraic function of z.

6. From 3, 4, 5 we obtain the following result. In order to obtain all actual points $\mathfrak{P}$ exactly once, one takes an arbitrary variable in z in the field Ω. If one constructs all the prime ideals $\mathfrak{p}$ in z and the null point for each, then one obtains all the points $\mathfrak{P}$ at which z remains finite. If $\mathfrak{P}'$ is a point different from these, then $z' = \frac{1}{z}$ has the finite value zero at $\mathfrak{P}'$. Conversely, each point $\mathfrak{P}'$ at which z' has the value zero is different from the points $\mathfrak{P}$. The prime ideal $\mathfrak{p}'$ in z' generated by such a point $\mathfrak{P}'$ (which consists of all integral algebraic functions of z' vanishing at $\mathfrak{P}'$) divides z', and conversely, the null point of all the prime ideals $\mathfrak{p}'$ in z' that divide z' is a point $\mathfrak{P}'$ at which $z' = 0$ and hence $z = \infty$. This finite number of extra points corresponding to the different $\mathfrak{p}'$, together with those previously derived from the prime ideals $\mathfrak{p}$ in z, exhausts the collection of all points $\mathfrak{P}$ making up the *Riemann surface T*.

§15. The order numbers

SUMMARY AND COMMENTS

This section sets up the concepts needed to discuss zeros and poles of algebraic functions. Zeros of a function ρ are described as points $\mathfrak{P}$ where ρ becomes "infinitely small," and the *order* of the zero is the highest power of the ideal $\mathfrak{p}$, generated by $\mathfrak{P}$, dividing the upper ideal of ρ.

Similarly, poles of a function η are points where η "becomes infinite," i.e., where $1/\eta$ becomes infinitely small. If this zero of $1/\eta$ has order r, then the pole of η is assigned order $-r$. At each point $\mathfrak{P}$, the assignment of order numbers to functions $\eta \in \Omega$ is an example of what we now call a *discrete valuation* on Ω.

The possibility of describing a function by its zeros and poles, and their orders, leads Dedekind and Weber to introduce formal

---

[38] If $\eta' = c'\rho^{m'} + \eta'_1\rho^{m'+1}$ then e.g.,

$$\frac{\eta}{\eta'} = \rho^{m-m'}\left(\frac{c}{c'} + \rho\eta''_1\right),$$

where

$$\eta''_1 = \frac{c'\eta_1 - c\eta'_1}{c'(c' + \rho\eta'_1)}$$

is a function of the same kind as η_1 (the proof is even simpler in the other cases).

products of points that they call *polygons*. Thus a polygon looks
like

$$\mathfrak{P}^r \mathfrak{P}_1^{r_1} \mathfrak{P}_2^{r_2} \cdots,$$

where $\mathfrak{P}, \mathfrak{P}_1 \mathfrak{P}_2, \ldots$ are points and $r, r_1, r_2, \ldots$ are positive integers. Today, these objects are called (positive) *divisors* and they are usually written additively:

$$r\mathfrak{P} + r_1 \mathfrak{P}_1 + r_2 \mathfrak{P}_2 + \cdots.$$

The multiplicative notation has the advantage that the *greatest common divisor* and *least common multiple* can be defined in the obvious way.

It is likewise the case that polygons have straightforward multiplication and divisibility. As Dedekind and Weber say, in subsection 6, "the laws of divisibility of polygons agree completely with those for integers and ideals."

Polygons are natural for the description of zeros, but not for poles, because the exponents $r, r_1, r_2, \ldots$ are required to be positive. This is in line with the privileged role of ideals in the Dedekind-Weber theory—a polygon $\mathfrak{A} = \mathfrak{P}^r \mathfrak{P}_1^{r_1} \mathfrak{P}_2^{r_2} \cdots$ generates the ideal $\mathfrak{a}$ of integral algebraic functions that vanish at $\mathfrak{P}, \mathfrak{P}_1, \mathfrak{P}_2, \ldots$ with orders $r, r_1, r_2, \ldots$ respectively—but some amendment is required to take poles into account. This is done in §17 by means of *quotients* $\mathfrak{A}/\mathfrak{B}$ of polygons.

Finally, this section takes the first step towards extracting the "surface" structure from the Riemann surface, which until now has been merely a structureless set of "points." For a field Ω of degree n relative to z (in which case z is said to be of order n) it is shown that z takes each value c at exactly n points. Thus the points of the Riemann surface for Ω arrange themselves into n "sheets" covering the "Riemann sphere" $\mathbb{C} \cup \{\infty\}$ of values c.

1. *Definition.* If $\mathfrak{P}$ is a particular point, we consider all functions π in Ω vanishing at $\mathfrak{p}$ and assign each of them an *order number* as follows.

Such a function ρ has order number 1, or is said to be *infinitely small of first order* or 0^1 at $\mathfrak{P}$ when *all* quotients $\frac{\pi}{\rho}$ remain finite at $\mathfrak{P}$. If ρ' is a function like ρ, then $\frac{\rho'}{\rho}$ is neither 0 nor ∞ at $\mathfrak{P}$. Conversely, if $\frac{\rho'}{\rho}$ is neither 0 nor ∞ at $\mathfrak{P}$ then ρ' is likewise infinitely small of first order. Moreover, if for any function π there is a positive integer exponent r such that $\frac{\pi}{\rho^r}$ is neither 0 nor ∞ at $\mathfrak{P}$, then the same holds for $\frac{\pi}{\rho'^r}$, and π receives the order number r, or is said to be infinitely small of order r at the point $\mathfrak{P}$. We shall also say that π is 0^r at $\mathfrak{P}$ or π is 0 at $\mathfrak{P}^r$.

In order to decide the question of the existence of such functions ρ and such order numbers r, one takes any variable z that is finite at $\mathfrak{P}$, lets $\mathfrak{p}$ denote the prime ideal in z generated by $\mathfrak{P}$, and represents each function π (with the exception of the orderless constant 0) by §12 as the quotient of two relatively prime ideals in z. The upper ideal of each of these functions is then divisible by $\mathfrak{p}$, and there will also be such functions whose upper ideal is not divisible by $\mathfrak{p}^2$. The latter have the order number 1. For the remaining functions π the order number is the exponent of the highest power of $\mathfrak{p}$ dividing the upper ideal, as follows immediately from the theorems of §12.

2. If a function η has the finite value η_0 at $\mathfrak{P}$ then we say that η has this value r-tuply at $\mathfrak{P}$ or at r points coinciding with $\mathfrak{P}$, or at $\mathfrak{P}^r$, when the function $\eta - \eta_0$ is infinitely small of order r at $\mathfrak{P}$. However, if $\eta_0 = \infty$ we say that η has the value ∞ r-tuply at $\mathfrak{P}$, or at r points coinciding with $\mathfrak{P}$, or that η is ∞^r at $\mathfrak{P}$ or ∞ at $\mathfrak{P}^r$, when $\frac{1}{\eta}$ vanishes at $\mathfrak{P}^r$.

3. If a function η becomes ∞^r at $\mathfrak{P}$ then we assign it the order number $-r$, but when η is neither 0 nor ∞ at $\mathfrak{P}$ it is assigned order number 0. It follows that each function in the field Ω has a definite *order number* at an arbitrary point $\mathfrak{P}$, with the exception of the two constant functions 0 and ∞.

4. If ρ is a function that has order number 1 at a point $\mathfrak{P}$, and if η is a function with the (positive, negative or vanishing) order number m, then the concluding theorem of §12 makes it possible to determine, for each positive r, a series of constants $c_0, c_1, \ldots, c_{r-1}$, the first of which does not vanish, and a function σ that is finite at $\mathfrak{P}$, so that

$$\eta = c_0 \rho^m + c_1 \rho^{m+1} + \cdots + c_{r-1} \rho^{m+r-1} + \sigma \rho^{m+r}.$$

5. It follows immediately that the order number of a product of two or more functions equals the sum of the order numbers of the individual factors.

The order number of a quotient of two functions equals the order number of the numerator minus that of the denominator.

If $\eta_1, \eta_2, \ldots, \eta_s$ is a sequence of functions and m is the algebraically smallest of their order numbers then

$$\eta_1 = e_1 \rho^m + \sigma_1 \rho^{m+1},$$
$$\eta_2 = e_2 \rho^m + \sigma_2 \rho^{m+1},$$
$$\cdots\cdots\cdots\cdots\cdots$$
$$\eta_s = e_s \rho^m + \sigma_s \rho^{m+1},$$

where the constants $e_1, e_2, \ldots, e_s$ are all nonvanishing. Hence, if $c_1, c_2, \ldots, c_s$ are constants, the order number of

$$\eta = c_1 \eta_1 + c_2 \eta_2 + \cdots + c_s \eta_s$$

is m if $c_1 e_1 + c_2 e_2 + \cdots + c_s e_s$ is nonzero; otherwise it is greater than m.

6. We give the name *polygons*[39] to complexes of points, which may contain the same point more than once, and denote them by $\mathfrak{A}, \mathfrak{B}, \mathfrak{C}, \ldots$.

We also let $\mathfrak{A}\mathfrak{B}$ denote the polygon obtained from the points of polygons $\mathfrak{A}$ and $\mathfrak{B}$ by letting a point $\mathfrak{P}$ that appears r-tuply in $\mathfrak{A}$ and s-tuply in $\mathfrak{B}$ appear $(r+s)$-tuply in $\mathfrak{A}\mathfrak{B}$. The meaning of $\mathfrak{P}^r$ and $\mathfrak{A} = \mathfrak{P}\mathfrak{P}_1^{r_1}\mathfrak{P}_2^{r_2}\cdots$ is then immediate, and the laws of divisibility of polygons agree completely with those for integers and ideals. Points play the role of prime factors. Thus in order to obtain the unit as well one must admit the polygon $\mathfrak{O}$ containing no points (the null-gon).

The number of points of a polygon is called its *order*. A polygon of order n is called an *n-gon* for short.

---

[39]Polygons are today called (positive) *divisors*. Dedekind and Weber do not consider arbitrary divisors, which may include points with negative integers. Instead they use *quotients* of polygons (see §17). (Translator's note.)

The *greatest common divisor* of two polygons $\mathfrak{A}, \mathfrak{B}$ is the polygon that contains each point the *least* number of times it occurs in $\mathfrak{A}$ or $\mathfrak{B}$. If it is $\mathfrak{O}$, then $\mathfrak{A}, \mathfrak{B}$ are called *relatively prime*.

The *least common multiple* of $\mathfrak{A}$ and $\mathfrak{B}$ is the polygon that contains each point the *greatest* number of times it occurs in $\mathfrak{A}$ or $\mathfrak{B}$. If $\mathfrak{A}, \mathfrak{B}$ are relatively prime then $\mathfrak{A}\mathfrak{B}$ is their least common multiple.

If $\mathfrak{A} = \mathfrak{P}^r \mathfrak{P}_1^{r_1} \mathfrak{P}_2^{r_2} \cdots$ is an arbitrary polygon, then there are always functions z in Ω that are infinite at none of the points of $\mathfrak{A}$. Because if z is infinite at some points of $\mathfrak{A}$, one can choose a constant c so that $z - c$ has the value 0 at no point of $\mathfrak{A}$, and then $\frac{1}{z-c}$ is finite at all points of the polygon $\mathfrak{A}$. If one begins with such a variable, then the collection of all those integral algebraic functions of z that vanish at the points of the polygon $\mathfrak{A}$ (each counted according to its multiplicity) forms an ideal $\mathfrak{a} = \mathfrak{p}^r \mathfrak{p}_1^{r_1} \mathfrak{p}_2^{r_2} \cdots$, and one can say that the polygon $\mathfrak{A}$ *generates* the ideal $\mathfrak{a}$, or that $\mathfrak{A}$ is the *null polygon* of the ideal $\mathfrak{a}$. Here the ideal concept agrees completely with the concept of a system of integral algebraic functions that all vanish at the same fixed points. The ideal $\mathfrak{o}$ is generated by the null-gon $\mathfrak{O}$.

The product of two or more ideals is generated by the product of the null polygons of the factors, the greatest common divisor and least common multiple by the greatest common divisor and least common multiple of the corresponding null polygons.

7. *Theorem.* If z is any variable in Ω and n is the degree of the field Ω relative to z, then z takes each particular value c at exactly n points. Because if $\mathfrak{o}$ denotes the system of all integral algebraic functions of z and c is a finite constant, then

$$\mathfrak{o}(z - c) = \mathfrak{p}_1^{e_1} \mathfrak{p}_2^{e_2} \cdots, \quad e_1 + e_2 + \cdots = n \quad (\S\,9, 7),$$

where $\mathfrak{p}_1, \mathfrak{p}_2, \ldots$ are distinct prime ideals in z. If one lets $\mathfrak{P}_1, \mathfrak{P}_2, \ldots$ be the null points of $\mathfrak{p}_1, \mathfrak{p}_2, \ldots$ then, by 2, z has the value c at e_1 points $\mathfrak{P}_1$ (or at $\mathfrak{P}_1^{e_1}$), at e_2 points $\mathfrak{P}_2$ (or at $\mathfrak{P}_2^{e_2}$), etc., hence at the n points of the polygon $\mathfrak{P}_1^{e_1} \mathfrak{P}_2^{e_2} \cdots$. Conversely: if $\mathfrak{P}$ is a point at which z has the value c, and if $\mathfrak{p}$ is the prime ideal in z generated by $\mathfrak{P}$, then $z \equiv c \pmod{\mathfrak{p}}$ and hence $\mathfrak{p}$ is one of the ideals $\mathfrak{p}_1, \mathfrak{p}_2, \ldots$, so $\mathfrak{P}$ is one of the points $\mathfrak{P}_1, \mathfrak{P}_2, \ldots$. The same result also holds for $c = \infty$, because n is also the degree of Ω relative to $\frac{1}{z}$, hence if the latter variable takes the value 0, z takes the value ∞, at exactly n points. It follows from §11 that there are only a finite number of values of the constant c for which any of the exponents $e_1, e_2, \ldots$ can be greater than 1.

The number n, i.e., the number of points at which the function z has the same value, is called the *order* of the function z. The constant functions, and only these, have order zero. For all other functions in Ω the order is a positive integer. The order of a variable z is at the same time the degree of the field Ω relative to z.

§16. Conjugate points and conjugate values

SUMMARY AND COMMENTS

After observing in §15 that there are n points of the Riemann surface where a function $z \in \Omega$ takes a given value $c \in \mathbb{C} \cup \{\infty\}$, Dedekind and Weber now call these points *conjugate* points for z. This recalls the concept of "conjugate" in algebraic number theory, where each algebraic number α satisfies an equation whose roots are called its conjugates, and whose norm $N(\alpha)$ is the product of these conjugates.

Here, the norm $N_z(\eta)$ of a function η with respect to the variable z—previously defined differently in §2—turns out to equal the product of the conjugates of η.

The other development in this section is the unveiling of the *ramification* points and their connection with the *ramification ideal* of §11. The ramification points are those n-tuples of conjugate points that are not all distinct, these being the points where one or more "sheets" of the Riemann surface come together. The ramification ideal is generated by the ramification points at which the independent variable z has a finite value.

Another concept that will be important later (from §22 onward) is the *ramification number*, which counts the number of ramification points. In this section the connection between ramification and the discriminant, revealed in §11, is strengthened by relating the ramification number to the degree of the discriminant.

1. *Definition.* If c is a particular numerical value then, as shown in §15, there is a corresponding polygon $\mathfrak{A}$ of n (not necessarily distinct) points $\mathfrak{P}', \mathfrak{P}'', \ldots, \mathfrak{P}^{(n)}$ at which the nth order variable z has this value. These n points will be called *conjugate* for z. Any one of them (together with z) determines the others. If one lets c continuously assume all possible values, then the polygon $\mathfrak{A} = \mathfrak{P}'\mathfrak{P}'' \cdots \mathfrak{P}^{(n)}$ moves, and indeed in such a way that all its points change simultaneously. One thereby obtains all existing points, with multiple points occurring (finitely often) only where $z - z_0$ or $\frac{1}{z}$ vanishes to an order higher than first. Hence the product of all these polygons,

$$\prod \mathfrak{A} = T \mathfrak{Z}_z,$$

where T is the collection of all simple points, is the *Riemann surface*; $\mathfrak{Z}_z$ is a particular finite polygon, called the *ramification* or *winding polygon* of T in z. Each point $\mathfrak{Q}$ in $\mathfrak{Z}_z$ is called a *ramification* or *winding point of T in z*, of order s when it appears s-tuply in $\mathfrak{Z}_z$. We have $s = e - 1$ when $z - z_0$ or $\frac{1}{z}$ is infinitely small of eth order at $\mathfrak{Q}$. The order of the polygon $\mathfrak{Z}_z$ is called the *ramification* or *winding number* w_z of the surface T for z. Those points of the ramification polygon at which z has a finite value together generate the *ramification ideal* in z (§11).

If one wishes to pass from this definition of the "absolute" Riemann surface, which is an invariant concept of the field, to the well-known conception of Riemann, then one has to think of the surface spread over a z-plane, which is then covered n-tuply everywhere except at the ramification points.

2. *Theorem.* If

$$z' = \frac{c + dz}{a + bz},$$

where a, b, c, d are constants whose determinant $ad - bc$ is nonzero,[40] then

$$\mathfrak{Z}_z = \mathfrak{Z}_{z'}; \quad w_z = w_{z'}.$$

---

[40]Recall that this is the concept of *linear* transformation discussed by Dedekind and Weber in §13, where it was shown that such transformations leave norm, trace, and discriminant invariant. (Translator's note.)

Because if $z - z_0$ or $\frac{1}{z}$ is infinitely small of eth order at a point $\mathfrak{P}$ then this is also the case for

$$z' - z_0' = \frac{(ad - bc)(z - z_0)}{(a + bz)(a + bz_0)},$$

or, when z_0 is infinite, for

$$z' - z_0' = \frac{-(ad - bc)}{b(a + bz)},$$

when $z_0' = \infty$ and hence $a + bz_0 = 0$, for

$$\frac{1}{z'} = \frac{a + bz}{c + dz}.$$

In particular, if $z' = \frac{1}{z}$ then the ramification number $w_z = w_{z'}$ is equal to the degree of the discriminant $\Delta_z(\Omega)$ minus the number of vanishing roots of $\Delta_{z'}(\Omega) = 0$ (§11).

3. *Definition.* The values $\eta', \eta'', \ldots, \eta^{(n)}$ taken by an arbitrary function η in Ω at n points $\mathfrak{P}', \mathfrak{P}'', \ldots, \mathfrak{P}^{(n)}$ that are conjugate for z are called *conjugate values of η for z*.

4. *Theorem.* If $N_z(\eta)$ is the norm of an arbitrary function η relative to z, then the value that this rational function of z has for $z = z_0$ is equal to the product $\eta'\eta'' \cdots \eta^{(n)}$ of the conjugate values of η belonging to $z = z_0$, leaving aside the case in which this product is indeterminate, i.e., where one of these conjugate values is 0 and another is ∞.

To prove this theorem we can assume that z_0 is finite; because if $z_0 = \infty$ we replace z by the variable $z' = \frac{1}{z}$, which leaves the norm unaltered. We can also assume that the values $\eta', \eta'', \ldots, \eta^{(n)}$ are all finite; because if one of them is infinite none of the others is zero, by hypothesis, and we consider the function $\frac{1}{\eta}$ instead of η.

Now, under these hypotheses, let

$$\mathfrak{o}(z - z_0) = \mathfrak{p}_1^{e_1} \mathfrak{p}_2^{e_2} \mathfrak{p}_3^{e_3} \cdots$$

and let $\mathfrak{P}_1, \mathfrak{P}_2, \mathfrak{P}_3, \ldots$ be the null points of the distinct prime ideals $\mathfrak{p}_1, \mathfrak{p}_2, \mathfrak{p}_3, \ldots$. We construct a system of integral algebraic functions λ, μ of z by the following rule. Let

λ_1 be divisible by $\mathfrak{p}_1$, not by $\mathfrak{p}_1^2$;

λ_2 be divisible by $\mathfrak{p}_2$, not by $\mathfrak{p}_2^2$;

λ_3 be divisible by $\mathfrak{p}_3$, not by $\mathfrak{p}_3^2$;

$$\ldots\ldots\ldots\ldots\ldots\ldots\ldots\ldots\ldots\ldots$$

μ_1 be divisible by $\mathfrak{p}_2^{e_2}, \mathfrak{p}_3^{e_3}, \ldots$, not by $\mathfrak{p}_1, \mathfrak{p}_2^{e_2+1}, \mathfrak{p}_3^{e_3+1}, \ldots$;

μ_2 be divisible by $\mathfrak{p}_1^{e_1}, \mathfrak{p}_3^{e_3}, \ldots$, not by $\mathfrak{p}_2, \mathfrak{p}_1^{e_1+1}, \mathfrak{p}_3^{e_3+1}, \ldots$;

μ_3 be divisible by $\mathfrak{p}_1^{e_1}, \mathfrak{p}_2^{e_2}, \ldots$, not by $\mathfrak{p}_3, \mathfrak{p}_1^{e_1+1}, \mathfrak{p}_2^{e_2+1}, \ldots$;[41]

$$\ldots\ldots\ldots\ldots\ldots\ldots\ldots\ldots\ldots\ldots\ldots\ldots\ldots\ldots$$

---

[41]Such functions can be determined, by §9, 3, footnote, or by §11, 2 where one can set, e.g., $\lambda = \rho - b, \mu\lambda^e = \psi(\rho)$.

The n functions

$$\begin{array}{ccccc} \mu_1, & \mu_1\lambda_1, & \mu_1\lambda_1^2, & \ldots & \mu_1\lambda_1^{e_1-1}, \\ \mu_2, & \mu_2\lambda_2, & \mu_2\lambda_2^2, & \ldots & \mu_2\lambda_2^{e_2-1}, \\ \mu_3, & \mu_3\lambda_3, & \mu_3\lambda_3^2, & \ldots & \mu_3\lambda_3^{e_3-1}, \\ \ldots & \ldots & \ldots & \ldots & \ldots \end{array}$$

which we denote by $\eta_1, \eta_2, \ldots, \eta_n$, then form a basis of Ω. This assertion is contained in the more general assertion about to be proved.

If

$$(z - z_0)\zeta = x_1\eta_1 + x_2\eta_2 + \cdots + x_n\eta_n$$

with polynomial coefficients $x_1, x_2, \ldots, x_n$, and if ζ has finite values $\zeta', \zeta'', \zeta''', \ldots$ at the points $\mathfrak{P}_1, \mathfrak{P}_2, \mathfrak{P}_3, \ldots$, then all the coefficients $x_1, x_2, \ldots, x_n$ must be divisible by $z - z_0$. In fact, e.g., the left-hand side is infinitely small of order at least e_1 at the point $\mathfrak{P}_1$. Hence by §15, 5

$$x_1\eta_1 + x_2\eta_2 + \cdots + x_{e_1}\eta_{e_1} = \mu_1(x_1 + x_2\lambda_1 + \cdots + x_{e_1}\lambda_1^{e_1-1})$$

must also be infinitely small of this order. But this is impossible unless $x_1, x_2, \ldots, x_{e_1}$ vanish at $\mathfrak{P}_1$, and hence are divisible by $z - z_0$. Q.E.D.

It follows that we can set

$$\begin{aligned} \eta\mu_1\lambda_1^r = \ & \mu_1(x_1^{(0)} + x_1^{(1)}\lambda_1 + \cdots + x_1^{(e_1-1)}\lambda_1^{e_1-1}) \\ & + \mu_2(x_2^{(0)} + x_2^{(1)}\lambda_2 + \cdots + x_2^{(e_2-1)}\lambda_2^{e_2-1}) \\ & + \mu_3(x_3^{(0)} + x_3^{(1)}\lambda_3 + \cdots + x_3^{(e_3-1)}\lambda_3^{e_3-1}) + \cdots, \end{aligned}$$

where the $x_1^{(0)}, x_1^{(1)}, \ldots, x_2^{(0)}, \ldots$ are rational functions of z that all remain finite for $z = z_0$. At the points $\mathfrak{P}_2, \mathfrak{P}_3, \ldots$ the left-hand side is infinitely small of order at least $e_2, e_3, \ldots$. The same holds for $\mu_1, \mu_3, \ldots$ at $\mathfrak{P}_2$, but not for μ_2, for $\mu_1, \mu_2, \ldots$ at $\mathfrak{P}_3$ but not for $\mu_3, \ldots$. Consequently

$$\begin{array}{ccccc} x_2^{(0)} = 0, & x_2^{(1)} = 0, & \ldots, & x_2^{(e_2-1)} = 0, \\ x_3^{(0)} = 0, & x_3^{(1)} = 0, & \ldots, & x_3^{(e_3-1)} = 0, \\ \ldots & \ldots & \ldots & \ldots \end{array}$$

for $z = z_0$. At $\mathfrak{P}_1$ the left-hand side is infinitely small of order at least r. Therefore, when $r < e_1$

$$x_1^{(0)} = 0, \quad x_1^{(1)} = 0, \quad \ldots, \quad x_1^{(r-1)} = 0, \quad x_1^{(r)} = \eta'$$

for $z = z_0$. The same argument may be applied to the functions $\eta\mu_2\lambda_2^r, \eta\mu_3\lambda_3^r, \ldots$. Thus if one sets

$$\begin{aligned} \eta\eta_1 &= x_{1,1}\eta_1 + x_{1,2}\eta_2 + \cdots + x_{1,n}\eta_n, \\ \eta\eta_2 &= x_{2,1}\eta_1 + x_{2,2}\eta_2 + \cdots + x_{2,n}\eta_n, \\ & \cdots\cdots\cdots\cdots\cdots\cdots\cdots\cdots\cdots \\ \eta\eta_n &= x_{n,1}\eta_1 + x_{n,2}\eta_2 + \cdots + x_{n,n}\eta_n, \end{aligned}$$

then all the terms to the left of the diagonal vanish in the determinant

$$N(\eta) = \sum \pm x_{1,1}x_{2,2}\cdots x_{n,n}$$

for $z = z_0$, while e_1 of the diagonal terms equal η', e_2 of them equal η'', e_3 equal η''', ... Thus for $z = z_0$

$$N(\eta) = \eta'^{e_1} \eta''^{e_2} \eta'''^{e_3} \cdots .$$

Q.E.D.

5. Since

$$Tr(\eta) = x_{1,1} + x_{2,2} + \cdots + x_{n,n}$$

by the definition of trace, the same considerations lead to the theorem:

$$Tr(\eta) = e_1 \eta' + e_2 \eta'' + e_3 \eta''' + \cdots$$

for $z = z_0$, though only under the hypothesis that the values $\eta', \eta'', \eta''', \ldots$ are finite.

For an arbitrary constant (or rational function of z) t, Theorem 4 yields

$$N(t - \eta_0) = (t - \eta')^{e_1} (t - \eta'')^{e_2} (t - \eta''')^{e_3} \cdots$$

for $z = z_0$, and then comparing coefficients of equal powers of t yields an expression in terms of conjugate values (symmetric functions) for each of these coefficients.

6. If $\eta_1, \eta_2, \ldots, \eta_n$ is a basis for Ω, then 5 immediately yields the value of the discriminant of this system for $z = z_0$:

$$\Delta_z(\eta_1, \eta_2, \ldots, \eta_n) = \left(\sum \pm \eta_1' \eta_2'' \cdots \eta_n^{(n)} \right)^2,$$

where $\eta_\iota', \eta_\iota'', \ldots, \eta_\iota^{(n)}$ denote all the conjugate values of η_ι belonging to $z = z_0$, not necessarily distinct but assumed finite.

§17. Representing the functions in the field Ω by polygon quotients

SUMMARY AND COMMENTS

A special case of the theorem in §15 about the values taken by a function of order n is that a function of order zero is constant. It follows that each function η is determined, up to a constant factor, by the orders of its zeros and poles (or, as Dedekind and Weber say, by its order numbers at each point). This is because, if η' is any other function with the same order numbers at each point, η/η' is of order zero and hence constant.

Thus Dedekind and Weber recognize that both positive and negative order numbers are needed to determine a function η, and so they introduce *polygon quotients* $\mathfrak{A}/\mathfrak{B}$. The *upper* polygon $\mathfrak{A}$ encodes the zeros of η, and the *lower* polygon $\mathfrak{B}$ encodes the poles of η.

A function η in the field Ω has a nonzero order number at only a finite number of points. The sum of all these order numbers is 0, and hence the sum of the positive order numbers equals the sum of the negative order numbers, and indeed it equals the order of the function η (§15). If the order numbers of a function η are known for each point $\mathfrak{P}$, then the function η is determined up to a constant factor. Because if η' has everywhere the same order number as η then $\frac{\eta}{\eta'}$ (by §15, 5) has order number zero everywhere and hence (by §15, 7) is a constant.

Thus if we construct a polygon $\mathfrak{A}$ in which each point where η has a positive order number is taken this order number of times, and a second polygon $\mathfrak{B}$ in which we similarly take the points at which η has a negative order number, then the polygons $\mathfrak{A}, \mathfrak{B}$ are of the same order, and indeed it is the order of the function η. Thus the function η is determined up to a constant factor by these polygons $\mathfrak{A}, \mathfrak{B}$. We symbolically set

$$\eta = \frac{\mathfrak{A}}{\mathfrak{B}}$$

and call $\mathfrak{A}$ the *upper polygon*, and $\mathfrak{B}$ the *lower polygon*, of the function η.[42]

The way we have set things up, the two polygons $\mathfrak{A}, \mathfrak{B}$ are relatively prime. However, it is convenient to extend the notation so that one may also admit common factors in $\mathfrak{A}, \mathfrak{B}$, which is done by setting

$$\frac{\mathfrak{M}\mathfrak{A}}{\mathfrak{M}\mathfrak{B}} = \frac{\mathfrak{A}}{\mathfrak{B}},$$

where $\mathfrak{M}$ denotes an arbitrary polygon. If

$$\eta = \frac{\mathfrak{A}}{\mathfrak{B}}$$

in this generalised notation, then a point $\mathfrak{P}$ at which η has order number m will occur m_1-tuply in $\mathfrak{A}$ and m_2-tuply in $\mathfrak{B}$, where $m_1 - m_2 = m$. It is still true that the order of $\mathfrak{A}$ equals the order of $\mathfrak{B}$, but they no longer equal the order of the function η.

This definition (and §15, 5) yields the following theorem immediately: if

$$\eta = \frac{\mathfrak{A}}{\mathfrak{B}}, \quad \eta' = \frac{\mathfrak{A}'}{\mathfrak{B}'},$$

then

$$\eta\eta' = \frac{\mathfrak{A}\mathfrak{A}'}{\mathfrak{B}\mathfrak{B}'}, \quad \frac{\eta}{\eta'} = \frac{\mathfrak{A}\mathfrak{B}'}{\mathfrak{B}\mathfrak{A}'}.$$

By §14, 5 a *function η' is an integral algebraic function of η if and only if each point in the lower polygon of η' also appears in the lower polygon of η.*

§18. Equivalent polygons and polygon classes

Summary and Comments

Not all pairs $\mathfrak{A}, \mathfrak{B}$ can occur in an equation $\eta = \frac{\mathfrak{A}}{\mathfrak{B}}$, and Weber (1908), §191, remarks: "Investigating this relationship is the great question that, in another way, is answered by Abel's theorem."

The present section begins the march towards Abel's theorem by defining equivalence classes of polygons, the product of equivalence classes, and divisibility of equivalence classes.

---

[42] All functions in the simple vector space (η) therefore have the same symbol $\frac{\mathfrak{A}}{\mathfrak{B}}$, and it would therefore be more correct to set $(\eta) = \frac{\mathfrak{A}}{\mathfrak{B}}$. However, the latter notation leads to unnecessary longwindedness.

1. *Definition.* Two polygons $\mathfrak{A}, \mathfrak{A}'$ with the same number of points are called *equivalent*[43] when there is a function η in Ω with the notation (§17)

$$\eta = \frac{\mathfrak{A}}{\mathfrak{A}'}.$$

2. *Theorem.* If $\mathfrak{A}$ is equivalent to $\mathfrak{A}'$ and $\mathfrak{A}''$, then $\mathfrak{A}'$ is also equivalent to $\mathfrak{A}''$. Because

$$\eta' = \frac{\mathfrak{A}'}{\mathfrak{A}}, \quad \eta'' = \frac{\mathfrak{A}''}{\mathfrak{A}}$$

imply

$$\frac{\eta'}{\eta''} = \frac{\mathfrak{A}'}{\mathfrak{A}''}.$$

3. *Definition and Theorem.* All the polygons $\mathfrak{A}', \mathfrak{A}'', \ldots$ equivalent to a given polygon $\mathfrak{A}$ constitute a *polygon class* A. By 2, each polygon belongs to one and *only* one class, because if $\mathfrak{A}, \mathfrak{B}$ are two equivalent polygons that lead to the classes A, B, then, by 2, each polygon in B also belongs to A and conversely, hence the two classes are identical.

All polygons in a class have the same order, which will be called the *order of the class*.

4. Nevertheless there exist polygons that are equivalent to no others, so each of them forms a class by itself. Such polygons may be called *isolated*.

5. If $\mathfrak{M}$ is an arbitrary polygon, and $\mathfrak{A}$ is equivalent to $\mathfrak{A}'$, then $\mathfrak{M}\mathfrak{A}$ is also equivalent to $\mathfrak{M}\mathfrak{A}'$. Conversely, the equivalence of $\mathfrak{M}\mathfrak{A}$ and $\mathfrak{M}\mathfrak{A}'$ implies the equivalence of $\mathfrak{A}$ and $\mathfrak{A}'$.

6. If $\mathfrak{A}$ is equivalent to $\mathfrak{A}'$, and $\mathfrak{B}$ to $\mathfrak{B}'$, then $\mathfrak{A}\mathfrak{B}$ is also equivalent to $\mathfrak{A}'\mathfrak{B}'$. The class C containing the product $\mathfrak{A}\mathfrak{B}$ therefore contains all products of two polygons respectively from the classes A, B of $\mathfrak{A}, \mathfrak{B}$ (as well as infinitely many other polygons under some circumstances), and we shall call it the *product* of the two classes A, B:

$$C = AB = BA.$$

The definition of the product of several classes, and the validity of the fundamental multiplication theorem,[44] are immediate.

7. If A, B, D are three classes that satisfy the condition

$$DA = DB,$$

then $A = B$ follows. Because if $\mathfrak{A}, \mathfrak{B}, \mathfrak{D}$ are three polygons in the classes A, B, D, then it follows from the hypothesis that $\mathfrak{D}\mathfrak{B}$ is equivalent to $\mathfrak{D}\mathfrak{A}$, and hence $\mathfrak{B}$ is equivalent to $\mathfrak{A}$.

8. If a polygon $\mathfrak{A}$ in class A divides a polygon $\mathfrak{C}$ in class C, then the same is true of each polygon $\mathfrak{A}'$ in class A. Because it follows from $\mathfrak{C} = \mathfrak{A}\mathfrak{B}$, by 5, that $\mathfrak{C}' = \mathfrak{A}'\mathfrak{B}$ is in C. We can therefore say, regardless of whether each polygon in class

---

[43]Today, this equivalence relation is called *linear equivalence*, and its equivalence classes are called *linear systems*. (Translator's note.)

[44]That is, associativity. (Translator's note.)

C is divisible by a polygon in class A, that *the class C is divisible by the class A.*
If $\mathfrak{B}'$ is any polygon in the class B of $\mathfrak{B}$, then $\mathfrak{C}'' = \mathfrak{A}'\mathfrak{B}'$ is also in C and hence

$$C = AB.$$

Thus if C is divisible by A then there is one and (by 7) *only* one class B that
satisfies the condition

$$C = AB.$$

§19. Vector spaces of polygons

SUMMARY AND COMMENTS

This section investigates vector spaces of polygons, bringing
the concepts of linear independence, basis, and dimension into play.
This leads, in §21, to the result that each equivalence class of polygons is a finite-dimensional vector space.

1. If $\mathfrak{A}_1, \mathfrak{A}_2, \ldots, \mathfrak{A}_s$ are particular equivalent polygons, and if $\mathfrak{A}$ is an arbitrary
polygon in the same class A, then there are s functions in Ω,

$$\eta_1 = \frac{\mathfrak{A}_1}{\mathfrak{A}}, \quad \eta_2 = \frac{\mathfrak{A}_2}{\mathfrak{A}}, \quad \ldots, \quad \eta_s = \frac{\mathfrak{A}_s}{\mathfrak{A}}.$$

If, at an arbitrary point $\mathfrak{P}$, one sets

$$\eta_1 = e_1 \rho^m + \sigma_1 \rho^{m+1},$$
$$\eta_2 = e_2 \rho^m + \sigma_2 \rho^{m+1},$$
$$\cdots\cdots\cdots\cdots\cdots$$
$$\eta_s = e_s \rho^m + \sigma_s \rho^{m+1},$$

as in §15, 5, where ρ is 0^1 at $\mathfrak{P}$, $e_1, e_2, \ldots, e_s$ are constants, not all zero, and
$\sigma_1, \sigma_2, \ldots, \sigma_s$ are functions finite at $\mathfrak{P}$, then it follows that each function η in the
vector space $(\eta_1, \eta_2, \ldots, \eta_s)$, i.e., each function of the form

$$\eta = c_1 \eta_1 + c_2 \eta_2 + \cdots + c_s \eta_s,$$

has an order number at $\mathfrak{P}$ no smaller than m, and hence, by §17, that the function
η can be expressed in the form

$$\eta = \frac{\mathfrak{A}'}{\mathfrak{A}},$$

where $\mathfrak{A}'$ likewise belongs to the class A.

If one replaces $\mathfrak{A}$ by any other polygon $\mathfrak{B}$ in the class A, and sets

$$\zeta = \frac{\mathfrak{A}}{\mathfrak{B}},$$

$$\eta_1 \zeta = \eta_1' = \frac{\mathfrak{A}_1}{\mathfrak{B}}, \quad \eta_2 \zeta = \eta_2' = \frac{\mathfrak{A}_2}{\mathfrak{B}}, \quad \ldots, \quad \eta_s \zeta = \eta_s' = \frac{\mathfrak{A}_s}{\mathfrak{B}},$$

then we also have

$$\eta \zeta = \eta' = c_1 \eta_1' + c_2 \eta_2' + \cdots + c_s \eta_s',$$

and hence

$$\eta' = \frac{\mathfrak{A}'}{\mathfrak{B}}.$$

Each polygon $\mathfrak{A}'$ generated by the denominator $\mathfrak{A}$ and the system $c_1, c_2, \ldots, c_s$ of constants is therefore also generated by any other denominator $\mathfrak{B}$ belonging to the same class, and the collection of all polygons $\mathfrak{A}'$ corresponding to the different values of the constants $c_1, c_2, \ldots, c_s$ is dependent only on the polygons $\mathfrak{A}_1, \mathfrak{A}_2, \ldots, \mathfrak{A}_s$. This collection will therefore be called a *vector space of polygons* with basis $\mathfrak{A}_1, \mathfrak{A}_2, \ldots, \mathfrak{A}_s$, and will be denoted by

$$(\mathfrak{A}_1, \mathfrak{A}_2, \ldots, \mathfrak{A}_s).$$

2. If the polygons $\mathfrak{A}_1, \mathfrak{A}_2, \ldots, \mathfrak{A}_s$ have a greatest common divisor $\mathfrak{M}$, then the latter is also a divisor of each polygon $\mathfrak{A}'$ in the vector space $(\mathfrak{A}_1, \mathfrak{A}_2, \ldots, \mathfrak{A}_s)$, by 1, and it may be called a *divisor*[45] *of the vector space*. However, it is possible to choose a polygon $\mathfrak{A}' = \mathfrak{M}\mathfrak{B}$ in this system so that $\mathfrak{B}$ is relatively prime to an arbitrary polygon. To show this, we retain the notation of 1 and suppose a point $\mathfrak{P}$ occurs μ-tuply in $\mathfrak{M}$ and ν-tuply in $\mathfrak{A}$. Then, if we set

$$\eta = e\rho^m + \sigma\rho^{m+1},$$

m is never less than $\mu - \nu$, and $m = \mu - \nu$ when one chooses the constants $c_1, c_2, \ldots, c_s$ so that

$$e = c_1 e_1 + c_2 e_2 + \cdots + c_s e_s$$

is nonzero. The point $\mathfrak{P}$ therefore occurs at least μ-tuply in $\mathfrak{A}'$ and, by the last hypothesis, no more than μ-tuply. Now since one can always choose the constants $c_1, c_2, \ldots, c_s$ so that an arbitrary number of expressions of the form

$$\sum c_\iota e_\iota, \quad \sum c_\iota e'_\iota, \quad \ldots,$$

in which not all the constants $e_\iota, e'_\iota, \ldots$ vanish, themselves have nonzero values, the correctness of the assertion follows.

3. The functions $\eta_1, \eta_2, \ldots, \eta_s$ in 1 are linearly independent if and only if the same is true of the functions $\eta'_1, \eta'_2, \ldots, \eta'_s$. We can therefore speak of the polygons $\mathfrak{A}_1, \mathfrak{A}_2, \ldots, \mathfrak{A}_s$ being *linearly dependent* or *independent*, and of their being a *linearly reducible* or *irreducible* system.

Since by §5, 4 each vector space of functions has an irreducible basis, it follows that each vector space of polygons has an *irreducible basis*. If s is the number of polygons in such a basis, then the vector space is said to be *s-tuple*, or of *dimension* s. Any s polygons in such a vector space constitute an irreducible basis if and only if they are independent (cf. §5, 4).

4. If $\mathfrak{M}$ is an arbitrary polygon, then the polygons $\mathfrak{A}_1, \mathfrak{A}_2, \ldots, \mathfrak{A}_s$ are linearly independent if and only if $\mathfrak{M}\mathfrak{A}_1, \mathfrak{M}\mathfrak{A}_2, \ldots, \mathfrak{M}\mathfrak{A}_s$ are linearly independent.

§20. Lowering the dimension of the space by divisibility conditions

Summary and Comments

By relating dimension to divisibility, Dedekind and Weber pave the way for bounding the dimension of an equivalence class of polygons in §21.

---

[45]Today called a *base locus*, while for us a "divisor" of the space is simply a member. (Translator's note.)

1. Let

$$S = (\mathfrak{A}_1, \mathfrak{A}_2, \ldots, \mathfrak{A}_s)$$

be an s-tuple vector space with divisor $\mathfrak{M}$. We consider the manifold of those polygons $\mathfrak{A}'$ in the system S that contain a given point at least once more often than the divisor $\mathfrak{M}$ of the system.

If point $\mathfrak{P}$ is contained μ-tuply in $\mathfrak{M}$ and ν-tuply in a polygon $\mathfrak{A}$ equivalent to $\mathfrak{A}_1, \mathfrak{A}_2$, then, when we set

$$\frac{\mathfrak{A}_1}{\mathfrak{A}} = \eta_1 = e_1\rho^m + \sigma_1\rho^{m+1},$$

$$\frac{\mathfrak{A}_2}{\mathfrak{A}} = \eta_2 = e_2\rho^m + \sigma_2\rho^{m+1},$$

$$\cdots\cdots\cdots\cdots\cdots\cdots\cdots\cdots$$

$$\frac{\mathfrak{A}_s}{\mathfrak{A}} = \eta_s = e_s\rho^m + \sigma_s\rho^{m+1},$$

$m = \mu - \nu$ and at least one of the constants $e_1, e_2, \ldots, e_s$, say e_s, is nonzero. The desired polygon $\mathfrak{A}'$ is then characterised by the equation

$$\frac{\mathfrak{A}'}{\mathfrak{A}} = \eta' = c_1\eta_1 + c_2\eta_2 + \cdots + c_s\eta_s,$$

where the constants $c_1, c_2, \ldots, c_s$ are related by the condition

$$c_1 e_1 + c_2 e_2 + \cdots + c_s e_s = 0.$$

It follows that we can set

$$\frac{\mathfrak{A}'}{\mathfrak{A}} = e_s\eta' = c_1(e_s\eta_1 - e_1\eta_2) + \cdots + c_{s-1}(e_s\eta_{s-1} - e_{s-1}\eta_s).$$

But then, when we set

$$\eta_1' = e_s\eta_1 - e_1\eta_s,$$

$$\eta_2' = e_s\eta_2 - e_2\eta_s,$$

$$\cdots\cdots\cdots\cdots\cdots\cdots$$

$$\eta_{s-1}' = e_s\eta_{s-1} - e_{s-1}\eta_s,$$

we see that the functions η' constitute an $(s-1)$-tuple vector space $(\eta_1', \eta_2', \ldots, \eta_{s-1}')$, because the functions $\eta_1', \eta_2', \ldots, \eta_{s-1}'$ are linearly independent when, as assumed here, the functions $\eta_1, \eta_2, \ldots, \eta_s$ are. Thus the polygons $\mathfrak{A}'$ also form an $(s-1)$-tuple vector space

$$S' = (\mathfrak{A}_1', \mathfrak{A}_2', \ldots, \mathfrak{A}_{s-1}')$$

when

$$\frac{\mathfrak{A}_\iota'}{\mathfrak{A}} = e_s\eta_\iota - e_\iota\eta_s.$$

The divisor of this vector space is divisible by $\mathfrak{M}\mathfrak{P}$, although not necessarily identical with it.

2. It follows immediately that the polygons of a vector space S that are divisible by an arbitrary r-gon $\mathfrak{R}$ form a vector space that is at *least* $(s-r)$-tuple. Because if we assume this already proved for an r-gon $\mathfrak{R}$, then the correctness of the assertion for an $(r+1)$-gon $\mathfrak{P}\mathfrak{R}$ follows immediately from 1, since the appearance of the point $\mathfrak{P}$, when $\mathfrak{P}$ is in the divisor of the vector space already reduced by $\mathfrak{R}$, cannot affect the dimension except to lower it by 1.

It follows in particular that an s-tuple vector space always contains *at least one* polygon divisible by a given $(s-1)$-gon.

3. When $r \leq s$ one can choose the r-gon $\mathfrak{R}$ in such a way that the polygons of the vector space S that are divisible by $\mathfrak{R}$ form a precisely $(s-r)$-tuple vector space. To do this one chooses a point $\mathfrak{P}$ not contained in the divisor of S. The polygons in S divisible by $\mathfrak{P}$ form an $(s-1)$-tuple vector space S' by 1. One chooses a second point $\mathfrak{P}'$ not in the divisor of S'. The polygons in S' divisible by $\mathfrak{P}'$, i.e., the polygons in S divisible by $\mathfrak{P}\mathfrak{P}'$, form an $(s-2)$-tuple vector space, etc. At the same time, it becomes clear from this construction that one can take $\mathfrak{R}$ to be relatively prime to an arbitrary given polygon. If $r = s$ this means that there is no polygon in S that is divisible by $\mathfrak{R}$.

§21. The dimensions of polygon classes

Summary and Comments

The study of equivalent polygons, begun in §18, now culminates in the result that a polygon equivalence class is finite-dimensional. The finite dimensionality makes it possible to define the concept of a *proper*[46] class; namely, a class is proper if the greatest common divisor of its basis elements is the empty polygon. This concept turns out to be useful in the proof of the Riemann-Roch theorem. The first, and most substantial, case to be proved is that for proper classes, in §28.

1. *The polygons in a class form a vector space of finite dimension, called the dimension of the class.*

Proof. If one chooses s polygons $\mathfrak{A}_1, \mathfrak{A}_2, \ldots, \mathfrak{A}_s$ in a class A of order m, then all polygons in the vector space $(\mathfrak{A}_1, \mathfrak{A}_2, \ldots, \mathfrak{A}_s)$ likewise belong to the class A. The number of linearly independent polygons in A can therefore not be greater than $m + 1$, otherwise (by §20, 2) one could find a polygon in the class divisible by an arbitrary $(m+1)$-gon, which is absurd. Thus if s is the maximal number of linearly independent polygons $\mathfrak{A}_1, \mathfrak{A}_2, \ldots, \mathfrak{A}_s$ in the class A then each polygon in the class must belong to the vector space $(\mathfrak{A}_1, \mathfrak{A}_2, \ldots, \mathfrak{A}_s)$, and s is the *dimension of the class*. The system $\mathfrak{A}_1, \mathfrak{A}_2, \ldots, \mathfrak{A}_s$ of polygons will be called a *basis of the class*.

The isolated polygons form classes of dimension 1.

2. If a class C contains s and no more linearly independent polygons divisible by a given polygon $\mathfrak{A}$ in the class A,

$$\mathfrak{C}_1 = \mathfrak{A}\mathfrak{B}_1, \quad \mathfrak{C}_2 = \mathfrak{A}\mathfrak{B}_2, \quad \ldots, \quad \mathfrak{C}_s = \mathfrak{A}\mathfrak{B}_s,$$

then C is divisible by A, and C also contains the same number of independent polygons

$$\mathfrak{C}_1' = \mathfrak{A}'\mathfrak{B}_1, \quad \mathfrak{C}_2' = \mathfrak{A}'\mathfrak{B}_2, \quad \ldots, \quad \mathfrak{C}_s' = \mathfrak{A}'\mathfrak{B}_s,$$

divisible by an arbitrary polygon $\mathfrak{A}'$ equivalent to $\mathfrak{A}$ (§18, 8; §19, 4). The number s therefore depends only on the classes A, C and can be conveniently denoted by (A, C). The value of the symbol (A, C) is set equal to 0 when C is not divisible by

---

[46]Today, a proper class is called linear system without base points. (Translator's note.)

A. It follows that the dimension of a class A is (O, A), where O denotes the class consisting of the null-gon $\mathfrak{O}$. If (§18, 8)

$$C = AB,$$

then it follows that

(1) $(A, C) = (A, AB) = (O, B),$

because the polygons $\mathfrak{B}_1, \mathfrak{B}_2, \ldots, \mathfrak{B}_s$, which all belong to B, are linearly independent, hence (O, B) is certainly not smaller than s. Conversely, if $\mathfrak{B}$ is an arbitrary polygon in class B, then $\mathfrak{AB}$ is in C, hence also in the vector space $(\mathfrak{AB}_1, \mathfrak{AB}_2, \ldots, \mathfrak{AB}_s)$, therefore $\mathfrak{B}$ is in the vector space $(\mathfrak{B}_1, \mathfrak{B}_2, \ldots, \mathfrak{B}_s)$, i.e., $(O, B) = s$.

If a is the order of the class A, then $(A, C) \geq (O, C) - a$, by §20, 2, and then by (1) we get the general theorem

(2) $(O, B) \geq (O, AB) - a.$

3. If all the basis polygons of a class A have the greatest common divisor $\mathfrak{M}$, then the latter is a divisor of all the polygons in class A. If $\mathfrak{M}$ equals the null-gon $\mathfrak{O}$, then the class is called *proper*, otherwise *improper with divisor* $\mathfrak{M}$.

If one divides all polygons in an improper class A by the divisor $\mathfrak{M}$, then one obtains a proper class A' of lower order, but of the same dimension. The relation between A and A' will be expressed symbolically by

$$A = \mathfrak{M}A'.$$

4. The divisor $\mathfrak{M}$ of an improper class A is always an isolated polygon. If

$$A = \mathfrak{M}A',$$

then by §19, 2 one can choose a polygon $\mathfrak{A}'$ in the proper class A' that is relatively prime to $\mathfrak{M}$. Thus if $\mathfrak{M}'$ is equivalent to $\mathfrak{M}$, then $\mathfrak{M}'\mathfrak{A}'$ is equivalent to $\mathfrak{M}\mathfrak{A}'$, and hence belongs to A, which means it is divisible by $\mathfrak{M}$. Therefore $\mathfrak{M}'$ is also divisible by $\mathfrak{M}$, and since $\mathfrak{M}$ and $\mathfrak{M}'$ are of the same order,

$$\mathfrak{M} = \mathfrak{M}'.$$

Consequently the single polygon $\mathfrak{M}$ constitutes a class M, and the symbol $\mathfrak{M}A'$ means the same as MA' (§18, 6).

§22. The normal bases of $\mathfrak{o}$

SUMMARY AND COMMENTS

A normal basis is one that suits not only the integral closure of $\mathbb{C}[z]$, but also the integral closure of $\mathbb{C}[z']$, where $z' = 1/z$, thus allowing equal treatment of ordinary points and points at infinity. The construction of normal bases is the technical heart of Dedekind and Weber's work, and it was reworked or rediscovered by many later mathematicians, among them Hensel and Landsberg, Birkhoff, Hasse, and Grothendieck. See Geyer (1981) for an account of these later developments and a modern formulation of the result in terms of vector bundles.

Geyer remarks, (p. 126), that the later proofs have "overall the same approach as Dedekind and Weber." Since Grothendieck is

famous for doing things his own way, this is strong evidence for the
"inner necessity" alluded to by Dedekind in his letter to Weber that
was quoted in Section 12 of the Translator's Introduction.

Some of the technical results on bases of o, harking back to
results in §2 and §3, throw light on the ramification number w_z
introduced in §16. Because of the relation between the ramification number and the degree of the discriminant, found in §16, the
ramification number is *even*. This is important in §24, where half
of the ramification number occurs in the definition of genus.

1. In what follows we consider the system o of integral algebraic functions ω of
an arbitrary variable z in Ω and at the same time the system o′ of integral algebraic
functions ω' of $z' = \frac{1}{z}$. It is immediately clear from the definition of integral
algebraic function that the two systems o, o′ have *only* the constants in common,
but that each function ω can be converted into a function ω' by multiplying it by
a particular positive power of z'. If $\omega z'^r$ is in o, then so too are $\omega z'^{r+1}, \omega z'^{r+2}, \ldots$.
Thus in the series of functions

$$\omega, \quad \frac{\omega}{z} = z'\omega, \quad \frac{\omega}{z^2} = z'^2\omega, \quad \ldots$$

there is a certain term $\omega z'^r$ that belongs to o′ along with all its successors, while
all the preceding terms do not. The smallest number r for which $z'^r\omega$ belongs to
o′ will be called the *exponent* of the function ω with respect to z. The constants,
and only these, have exponent zero. If ω is nonzero, and if r is its exponent, then
$r + 1$ is the exponent of $(z - c)\omega$. Because if $\omega = z'^r\omega'$, then

$$\frac{(z - c)\omega}{z^{r+1}} = (1 - cz')\omega' \quad \text{is in o}',$$

$$\frac{(z - c)\omega}{z^r} = z\omega' - c\omega' \quad \text{is not in o}',$$

since $c\omega'$ is in o′ but $z\omega' = \frac{\omega}{z^{r-1}}$ is not. It follows from this that, in general:

If x is a polynomial in z of degree s, and if r is the exponent of ω, then $(r + s)$
is the exponent of $x\omega$.

2. We now choose a function system $\lambda_1, \lambda_2, \ldots, \lambda_n$ in o by the following rule:

Let λ_1 be a nonzero constant, e.g., 1. Let λ_2 be among the functions in o,
not congruent to a constant modulo oz, that have smallest possible exponent r_2,
etc. In general, let λ_s be among the functions in o, not congruent to a function in
the vector space $(\lambda_1, \lambda_2, \ldots, \lambda_{s-1})$ (mod oz), and of smallest possible exponent r_s.
Since $(o, oz) = N(z) = z^n$ is of nth degree, there are exactly n linearly indepen-
dent functions modulo oz in o (§6), and hence the series of functions $\lambda_1, \lambda_2, \lambda_3, \ldots$
contains no more, and no less, than n terms. We then have (§5)

$$o \equiv (\lambda_1, \lambda_2, \ldots, \lambda_n) \quad (\text{mod } oz).$$

The exponents $r_1, r_2, \ldots, r_n$ of the functions $\lambda_1, \lambda_2, \ldots, \lambda_n$ satisfy the conditions

$$r_1 = 0, \quad 1 \le r_2 \le r_3 \le \cdots \le r_n.$$

Each function in o with exponent $< r_s$ is congruent modulo oz to a function in the
$(s - 1)$-tuple vector space

$$(\lambda_1, \lambda_2, \ldots, \lambda_{s-1}).$$

These functions $\lambda_1, \lambda_2, \ldots, \lambda_n$ *form a basis of* $\mathfrak{o}$, as we see from the following considerations.

If this were not the case, one could determine (§3, 7) a linear function $z - c$ and a system of constants $a_1, a_2, \ldots, a_n$, not all zero, such that

$$a_1 \lambda_1 + a_2 \lambda_2 + \cdots + a_n \lambda_n = (z - c)\omega.$$

If a_s is the last nonvanishing constant a, then

$$a_1 \lambda_1 + a_2 \lambda_2 + \cdots + a_s \lambda_s = (z - c)\omega,$$

and the exponent of ω is certainly less than r_s (since $\frac{(z-c)\omega}{z^{r_s}}$ is in $\mathfrak{o}'$). Thus ω is congruent to a function of the vector space $(\lambda_1, \lambda_2, \ldots, \lambda_{s-1})$ (mod $\mathfrak{o}z$), and then, since a_s is nonzero, so too is λ_s—contrary to hypothesis.

The functions $\lambda_1, \lambda_2, \ldots, \lambda_n$ therefore form a basis of $\mathfrak{o}$, which will be called a *normal basis*. The characteristic properties of normal bases are:

I. The functions $\lambda_1, \lambda_2, \ldots, \lambda_n$ are linearly independent modulo $\mathfrak{o}z$.

II. Each function in $\mathfrak{o}$ whose exponent is smaller than the exponent r_s of λ_s belongs to the form

$$c_1 \lambda_1 + c_2 \lambda_2 + \cdots + c_{s-1} \lambda_{s-1} + z\omega_s,$$

where $c_1, c_2, \ldots, c_{s-1}$ are constants and ω_s is a function in $\mathfrak{o}$.

3. The functions

$$\lambda_1' = \frac{\lambda_1}{z^{r_1}}, \quad \lambda_2' = \frac{\lambda_2}{z^{r_2}}, \quad \ldots, \quad \lambda_n' = \frac{\lambda_n}{z^{r_n}}$$

in $\mathfrak{o}'$ form a *normal basis* of $\mathfrak{o}'$.

If ω is a function in $\mathfrak{o}$ *not divisible by* z and of exponent r, then the exponent of $\omega' = \frac{\omega}{z^r}$ is likewise r with respect to z', because indeed $\frac{\omega'}{z'^r} = \omega$ is in $\mathfrak{o}$ but $\frac{\omega'}{z'^{r-1}} = \frac{\omega}{z}$ is not. Since the functions $\lambda_1, \lambda_2, \ldots, \lambda_n$ are all not divisible by z, it follows that the exponents of $\lambda_1', \lambda_2', \ldots, \lambda_n'$ with respect to z' are $r_1, r_2, \ldots, r_n$ respectively. Having established this, we now prove that the function system $\lambda_1', \lambda_2', \ldots, \lambda_n'$ has the properties I, II when the $\mathfrak{o}, z$ therein are replaced by $\mathfrak{o}', z'$.

If condition I is not satisfied then constants $a_1, a_2, \ldots, a_s$ may be determined, with the last one nonvanishing, so that

$$a_1 \lambda_1' + a_2 \lambda_2' + \cdots + a_s \lambda_s' = z'\omega',$$

and hence also (multiplying by z^{r_s})

$$a_1 z^{r_s - r_1} \lambda_1 + a_2 z^{r_s - r_2} \lambda_2 + \cdots + a_s \lambda_s = \omega,$$

whence

$$\omega = z^{r_s - 1} \omega',$$

is a function in $\mathfrak{o}$ with exponent less than r_s. Since a_s is nonzero, however, this is impossible by the hypothesis on λ. Consequently, condition I is satisfied, whence it follows that

$$\mathfrak{o}' \equiv (\lambda_1', \lambda_2', \ldots, \lambda_n') \pmod{\mathfrak{o}'z'}.$$

If condition II is not satisfied, and λ' is a function in $\mathfrak{o}'$ with exponent $r < r_s$, and not belonging to the form

$$a_1 \lambda_1' + a_2 \lambda_2' + \cdots + a_{s-1} \lambda_{s-1}' + z'\omega',$$

then one can choose $e \geq s$ so that

$$\lambda' = a_1 \lambda'_1 + a_2 \lambda'_2 + \cdots + a_e \lambda'_e + z' \omega'$$

with constant coefficients, the last of which, a_e, does not vanish. Consequently, we also have $r_e \geq r_s > r$.

This means that $\lambda = z^{r_e - 1} \lambda'$ is a function in $\mathfrak{o}$, and it follows by multiplication by z^{r_e} that

$$z\lambda = a_1 z^{r_e - r_1} \lambda_1 + a_2 z^{r_e - r_2} \lambda_2 + \cdots + a_e \lambda_e + z^{r_e - 1} \omega'.$$

Therefore $\omega = z^{r_e - 1} \omega'$ is a function in $\mathfrak{o}$ whose exponent (by 1) is $\leq r_e - 1$, and which satisfies the congruence

$$\omega \equiv a'_1 \lambda_1 + a'_2 \lambda_2 + \cdots + a'_e \lambda_e \pmod{\mathfrak{o}z},$$

where $a'_e = -a_e$ is nonzero. But then, by property II of the functions λ, we must have the exponent of $\omega \geq r_e$, which is clearly a contradiction.

This completes the proof that the function system $\lambda'_1, \lambda'_2, \ldots, \lambda'_n$ is a normal basis of $\mathfrak{o}'$.

4. We now construct the discriminant of Ω with respect to the variables z and z' with the help of the normal bases λ, λ'. We have

$$\Delta_z(\Omega) = \text{const.} \Delta(\lambda_1, \lambda_2, \ldots, \lambda_n),$$
$$\Delta_{z'}(\Omega) = \text{const.} \Delta(\lambda'_1, \lambda'_2, \ldots, \lambda'_n).$$

But if one replaces λ'_ι by the expression $z'^{r_\iota} \lambda_\iota$, then it follows from the theorem §2, (13) that

$$\Delta_{z'}(\Omega) = \text{const.} z'^{2(r_1 + r_2 + \cdots + r_n)} \Delta_z(\Omega).$$

If $\Delta_z(\Omega)$ is of degree δ, then $\Delta_{z'}(\Omega)$ has the $(2(r_1 + r_2 + \cdots + r_n) - \delta)$-tuple root $z' = 0$, and §16, 2 then gives the *ramification number*

$$w_z = 2(r_1 + r_2 + \cdots + r_n),$$

which is therefore always an *even number*.

§23. The differential quotient

SUMMARY AND COMMENTS

Since the "Riemann surfaces" of Dedekind and Weber lack the structure needed to perform integration, theorems that were originally about integrals (such as Abel's theorem and the residue theorem) are reformulated as theorems about differentials: the things that one integrates. Differentials can be defined algebraically, via the differential quotient.

But likewise, as Weber (1908), §194, remarks: "The differential quotient cannot be introduced in the normal way when we make no use of continuity." This is why the differential quotient $\frac{d\alpha}{d\beta}$ is approached in a somewhat roundabout way for functions $\alpha, \beta \in \Omega$. It turns out, however, that if $F(\alpha, \beta) = 0$ is the polynomial equation satisfied by α and β then

$$\frac{d\alpha}{d\beta} = -\frac{F'(\beta)}{F'(\alpha)}$$

as in traditional calculus, where Dedekind and Weber use $F'(\beta)$ to denote $\partial F/\partial \beta$ and $F'(\alpha)$ to denote $\partial F/\partial \alpha$. Since F is a polynomial, $\partial F/\partial \beta$ and $\partial F/\partial \alpha$ are also polynomials. Moreover, they have a simple algebraic meaning; for example $\partial F/\partial \alpha$ may be defined as the coefficient of δ in the polynomial $F(\alpha + \delta, \beta)$.

Likewise, the Taylor series for $F(\alpha, \beta)$, deployed in subsection 2, is actually a finite sum of polynomials.

We also have the chain rule

$$\frac{d\alpha}{d\beta} = \frac{d\alpha}{d\gamma}\frac{d\gamma}{d\beta},$$

and this makes it possible to associate a *differential* $d\alpha$ with each function α so that $\frac{d\alpha}{d\beta}$ is the quotient of the differentials $d\alpha$, $d\beta$. Moreover, we have the standard rules for the differential of the sum, difference, product, and quotient.

If ω is an integral algebraic function, then $\frac{d\omega}{dz}$ is not necessarily an integral algebraic function. However, it is shown in subsection 5 that $\frac{d\omega}{dz}$ belongs to the complementary module $\mathfrak{e}$ of the module $\mathfrak{o}$ of integral algebraic functions, introduced in §10. The ideal $\mathfrak{f}$ from §10 also returns in this section. In subsection 6 it turns out to be the "ideal of double points."

1.   Since each nonzero function in the field Ω takes the value zero at only a finite number of points, it follows that a function in Ω with infinitely many zeros is necessarily identically zero, and that two functions in Ω agreeing at infinitely many points are identical.

2.   If α, β are any two variables in the field Ω then there is a function in Ω, denoted by $\left(\frac{d\alpha}{d\beta}\right)$, which at infinitely many points $\mathfrak{P}$ satisfies the condition

$$\left(\frac{d\alpha}{d\beta}\right)_0 = \left(\frac{\alpha - \alpha_0}{\beta - \beta_0}\right)_0,$$

and is called the *differential quotient* of α with respect to β. Namely, if $F(\alpha, \beta) = 0$ is the irreducible equation relating α, β then, when we exclude the (finitely many) points at which α_0 or $\beta_0 = \infty$ or $F'(\alpha_0) = 0$ or $F'(\beta_0) = 0$, we obtain

$$0 = F(\alpha, \beta)$$
$$= F(\alpha_0, \beta_0) + (\alpha - \alpha_0)F'(\alpha_0) + (\beta - \beta_0)F'(\beta_0)$$
$$+ \frac{1}{2}\left\{(\alpha - \alpha_0)^2 F''(\alpha_0, \alpha_0) + 2(\alpha - \alpha_0)(\beta - \beta_0)F''(\alpha_0, \beta_0) + (\beta - \beta_0)^2 F''(\beta_0, \beta_0)\right\}$$
$$+ \cdots.$$

One of the two quotients $\left(\frac{\alpha-\alpha_0}{\beta-\beta_0}\right), \left(\frac{\beta-\beta_0}{\alpha-\alpha_0}\right)$ is certainly finite. If it is the first, then the equation above yields the following:

$$0 = \frac{\alpha-\alpha_0}{\beta-\beta_0} F'(\alpha_0) + F'(\beta_0)$$

$$+ (\beta-\beta_0)\frac{1}{2}\left\{\left(\frac{\alpha-\alpha_0}{\beta-\beta_0}\right)^2 F''(\alpha_0,\alpha_0) + 2\left(\frac{\alpha-\alpha_0}{\beta-\beta_0}\right) F''(\alpha_0,\beta_0) + F''(\beta_0,\beta_0)\right\}$$

$$+ \cdots,$$

whence it follows that at the points $\mathfrak{P}$:

$$\left(\frac{\alpha-\alpha_0}{\beta-\beta_0}\right)_0 = -\frac{F'(\beta_0)}{F'(\alpha_0)} = -\left(\frac{F'(\beta)}{F'(\alpha)}\right)_0.$$

Thus

$$(1) \qquad\qquad \left(\frac{d\alpha}{d\beta}\right) = -\frac{F'(\beta)}{F'(\alpha)}$$

has the desired property. This remains true when one of the functions α, β is a constant. Because if, e.g., α is constant, then $F(\alpha,\beta) = \alpha - \alpha_0$ is independent of β, so $F'(\alpha) = 1$, $F'(\beta) = 0$.

3. It follows from the foregoing that, when β is not constant, $\left(\frac{\alpha-\alpha_0}{\beta-\beta_0}\right)$ has a finite value except at a finite number of points. Therefore, if γ is a third variable in Ω, then

$$\left(\frac{\alpha-\alpha_0}{\beta-\beta_0}\right) = \left(\frac{\alpha-\alpha_0}{\gamma-\gamma_0}\right)\left(\frac{\gamma-\gamma_0}{\beta-\beta_0}\right)$$

at infinitely many points, so we also have

$$\left(\frac{d\alpha}{d\beta}\right)_0 = \left(\frac{d\alpha}{d\gamma}\right)_0\left(\frac{d\gamma}{d\beta}\right)_0.$$

From this and 1 we get the identity

$$(2) \qquad\qquad \left(\frac{d\alpha}{d\beta}\right) = \left(\frac{d\alpha}{d\gamma}\right)\left(\frac{d\gamma}{d\beta}\right). \quad [47]$$

4. As a consequence of the latter theorem we can associate with each function $\alpha, \beta, \gamma, \ldots$ in the field Ω a function $d\alpha, d\beta, d\gamma, \ldots$ (differential) in such a way that in general

$$\frac{d\alpha}{d\beta} = \left(\frac{d\alpha}{d\beta}\right).$$

---

[47]One can also define the differential quotient by the equation

$$\left(\frac{d\alpha}{d\beta}\right) = -\frac{F'(\beta)}{F'(\alpha)},$$

and prove the theorem

$$\left(\frac{d\alpha}{d\beta}\right) = \left(\frac{d\alpha}{d\gamma}\right)\left(\frac{d\gamma}{d\beta}\right)$$

by algebraic division.

The differentials of constants, and *only* these, are set equal to zero. The rest are completely determined as soon as one of them is taken arbitrarily. If the variables $\alpha, \beta, \gamma, \ldots$ satisfy a rational equation

$$F(\alpha, \beta, \gamma, \ldots) = 0,$$

then it follows that

(3) $$F'(\alpha)d\alpha + F'(\beta)d\beta + F'(\gamma)d\gamma + \cdots = 0.$$

Because one can conclude, as with (2), that this equation holds at infinitely many points.

Immediate consequences of the last theorem are the well-known rules for differentiation of sums, differences, products and quotients:

(4) $$d(\alpha \pm \beta) = d\alpha \pm d\beta,$$

(5) $$d(\alpha\beta) = \alpha\,d\beta + \beta\,d\alpha,$$

(6) $$d\left(\frac{\alpha}{\beta}\right) = \frac{\beta\,d\alpha - \alpha\,d\beta}{\beta^2}.$$

5. If ω is an *integral algebraic* function of z then $\frac{d\omega}{dz}$ is not in general an integral algebraic function of z. However, it is clear from the expression (§3, 7)

$$\omega = x_1\omega_1 + x_2\omega_2 + \cdots + x_n\omega_n,$$

since the differential quotients of the polynomial functions $x_1, x_2, \ldots, x_n$ are again polynomials, that the lower ideals of all the functions $\frac{d\omega}{dz}$ must divide a particular ideal, namely the least common multiple of the lower ideals of $\frac{d\omega_1}{dz}, \frac{d\omega_2}{dz}, \ldots, \frac{d\omega_n}{dz}$. We wish to find out what this ideal is. To that end let $z - c$ be an arbitrary linear function of z and let

$$\mathfrak{o}(z - c) = \mathfrak{p}^e\mathfrak{p}_1^{e_1}\mathfrak{p}_2^{e_2}\cdots,$$

where the prime ideals $\mathfrak{p}, \mathfrak{p}_1, \mathfrak{p}_2, \ldots$ are distinct. Now let ζ be the same function as in §11, 2, i.e., an integral algebraic function of z with distinct values at the points $\mathfrak{P}, \mathfrak{P}_1, \mathfrak{P}_2, \ldots$ generated by the prime ideals $\mathfrak{p}, \mathfrak{p}_1, \mathfrak{p}_2, \ldots$, and which takes each of these values simply. Then ω may be expressed in the form

$$\omega = y_0 + y_1\zeta + \cdots + y_{n-1}\zeta^{n-1},$$

where the rational functions $y_0, y_1, \ldots, y_{n-1}$ of z may be fractional, but the factor $z - c$ certainly does not occur in a denominator. It follows that the lower ideal of $\frac{d\omega}{dz}$ cannot be divisible by higher powers of the ideals $\mathfrak{p}, \mathfrak{p}_1, \mathfrak{p}_2, \ldots$ than the lower ideal of $\frac{d\zeta}{dz}$. But if the irreducible equation between ζ and z is

$$f(\zeta, z) = 0$$

then, by §11, 2,

$$\mathfrak{o}f'(\zeta) = \mathfrak{m}\mathfrak{p}^{e-1}\mathfrak{p}_1^{e_1-1}\mathfrak{p}_2^{e_2-1}\cdots$$

and $\mathfrak{m}$ is relatively prime to $\mathfrak{p}, \mathfrak{p}_1, \mathfrak{p}_2, \ldots$. However, since

$$\frac{d\zeta}{dz} = -\frac{f'(z)}{f'(\zeta)},$$

the lower ideal of $\frac{d\zeta}{dz}$, and hence also that of $\frac{d\omega}{dz}$, can contain none of the factors $\mathfrak{p}, \mathfrak{p}_1, \mathfrak{p}_2, \ldots$ more often than $(e - 1), (e_1 - 1), (e_2 - 1), \ldots$ times respectively. Now since $z - c$ can be an arbitrary linear function, it follows that $\frac{d\omega}{dz}$ can have no lower ideal but one that divides the ramification ideal $\mathfrak{z} = \prod \mathfrak{p}^{e-1}$ (§11). Thus we have

$$\mathfrak{z}\frac{d\omega}{dz} = \mathfrak{a},$$

where $\mathfrak{a}$ denotes an ideal, and hence by §11, (7)

$$\mathfrak{o}\frac{d\omega}{dz} = \mathfrak{e}\mathfrak{a},$$

whence it follows that the functions $\frac{d\omega}{dz}$ all belong to the complementary module $\mathfrak{e}$ of $\mathfrak{o}$.

6. If the irreducible equation $F(\omega, z) = 0$ between ω and z is of nth degree with respect to ω, so that $1, \omega, \omega^2, \ldots, \omega^{n-1}$ is a basis of Ω, then by §11, (10)

$$\mathfrak{o}F'(\omega) = \mathfrak{z}\mathfrak{f}.$$

Therefore, since,

$$\frac{d\omega}{dz} = -\frac{F'(z)}{F'(\omega)},$$

$\mathfrak{o}F'(z)$ must be divisible by the ideal $\mathfrak{f}$:

$$\mathfrak{o}F'(z) = \mathfrak{f}\mathfrak{a}.$$

One may therefore call $\mathfrak{f}$ the *ideal of the double points* with respect to ω, z.

7. If $\mathfrak{P}$ is a point at which $z - c$ is infinitely small of first order (and hence not a ramification point in z), then the functions $\frac{d\omega}{dz}$ are all finite at z, by 5. Thus if η is any function in Ω that is finite at $\mathfrak{P}$ then one can represent it as the quotient $\frac{\alpha}{\beta}$ of two integral algebraic functions, of which β does not vanish at $\mathfrak{P}$, so that $\frac{d\eta}{dz}$ is also finite at $\mathfrak{P}$ by 6.

8. Now let α, β be any two variables in Ω. We shall investigate the behaviour of $\frac{d\alpha}{d\beta}$ at any point $\mathfrak{P}$.

One chooses a variable z in Ω that is infinitely small of first order at $\mathfrak{P}$. If α has a finite value α_0 at $\mathfrak{P}$, then by §15, 1, and 2 one can determine a positive integer r and a function α' finite and nonzero at $\mathfrak{P}$ so that

$$\alpha = \alpha_0 + z^r\alpha'.$$

This still holds when α is infinite at $\mathfrak{P}$, except that r is then a negative integer and α_0 has to be replaced by an arbitrary finite constant, e.g., 0. Similarly one can set

$$\beta = \beta_0 + z^s\beta'.$$

Then r and s are the order numbers of $\alpha - \alpha_0$, $\beta - \beta_0$ at the point $\mathfrak{P}$, which may be positive or negative, but not zero. It then follows from (2) that

$$\frac{d\alpha}{d\beta} = z^{r-s}\frac{r\alpha' + z\frac{d\alpha'}{dz}}{s\beta' + z\frac{d\beta'}{dz}}$$

or

$$\frac{\beta - \beta_0}{\alpha - \alpha_0}\frac{d\alpha}{d\beta} = \frac{r + z\frac{d\alpha'}{\alpha' dz}}{s + z\frac{d\beta'}{\beta' dz}}.$$

Now if we again use the subscript 0 to denote the value of a function at the point $\mathfrak{P}$ then, since

$$\left(\frac{d\alpha'}{\alpha' dz}\right)_0, \quad \left(\frac{d\beta'}{\beta' dz}\right)_0$$

are finite by 7,

(7)
$$\left(\frac{\beta - \beta_0}{\alpha - \alpha_0}\frac{d\alpha}{d\beta}\right)_0 = \frac{r}{s}$$

is therefore finite and nonzero. It follows that *the order number of the differential quotient $\frac{d\alpha}{d\beta}$ is equal to the difference of the order numbers of $\alpha - \alpha_0$ and $\beta - \beta_0$.* If $r \neq s$ then $\left(\frac{\alpha - \alpha_0}{\beta - \beta_0}\right)_0$ and hence $\left(\frac{d\alpha}{d\beta}\right)_0$ is zero or infinite. On the other hand, if $r = s$ then both values are finite and nonzero, and hence we have, in all cases,

(8)
$$\left(\frac{\alpha - \alpha_0}{\beta - \beta_0}\right)_0 = \left(\frac{d\alpha}{d\beta}\right)_0.$$

Here, α_0 and β_0 are the values of α and β at $\mathfrak{P}$, when these values are finite; otherwise they are arbitrary constants, e.g., 0.

9. If a, b are the order numbers of $\alpha - \alpha_0$, $\beta - \beta_0$ at $\mathfrak{P}$ then, when a, b are positive, the point $\mathfrak{P}$ occurs $(a - 1)$-tuply, respectively, $(b - 1)$-tuply, in the ramification polygons $\mathfrak{Z}_\alpha, \mathfrak{Z}_\beta$ in α, β. But, if a is negative, $\mathfrak{Z}_a$ contains the point $\mathfrak{P}$ $(-a - 1)$-tuply, and similarly when b is negative (§16, 1). Thus if $\mathfrak{A}, \mathfrak{B}$ denote the lower polygons of α, β one obtains the following expression for $\frac{d\alpha}{d\beta}$ as a polygon quotient:

(9)
$$\frac{d\alpha}{d\beta} = \frac{\mathfrak{Z}_\alpha \mathfrak{B}^2}{\mathfrak{Z}_\beta \mathfrak{A}^2},$$

since the order number of $\frac{d\alpha}{d\beta}$ (as proved above) is always $a - b$.

§24. The genus of the field Ω

Summary and Comments

It follows from the Riemann-Hurwitz formula, discussed in Section 4 of the Translator's Introduction, that the genus p satisfies

$$p = \frac{1}{2}w - n + 1,$$

where w is the sum of the ramification numbers and n is the degree or "sheet number." As we said there, this motivates an algebraic definition of genus, since n and w are so definable.

Now is the time to introduce this definition, because the results of the previous section enable Dedekind and Weber to prove that $w_\alpha - 2n_\alpha$ is independent of the choice of function α, and hence that p is an invariant of the field Ω.

The remainder of the section investigates the relation between the genus p of Ω and the orders m and n of a pair of generating functions of Ω, ending with the formula

$$p = (n - 1)(m - 1) - r,$$

where r is a number that counts "double points."

1. If one lets w_α, w_β denote the ramification numbers, and n_α, n_β the orders, of the variables α, β, then it follows from formula (9) of the preceding section that, since numerator and denominator of $\frac{d\alpha}{d\beta}$ must have equal numbers of points, we have the important relation

$$w_\alpha - 2n_\alpha = w_\beta - 2n_\beta.$$

Thus when one sets

(1) $$p = \frac{1}{2}w - n + 1,$$

which is an integer by §22, 4, then this number is independent of the choice of variables. It is a characteristic number of the field Ω, called the *genus* of Ω. One sees that this number is never negative by replacing $\frac{1}{2}w$ by the value $r_1 + r_2 + \cdots + r_n$ from §22. Then one obtains

(2) $$p = (r_2 - 1) + (r_3 - 1) + \cdots + (r_n - 1),$$

which cannot be negative because $r_2, r_3, \ldots, r_n \geq 1$.

2. Let α, β be two functions in Ω with orders m, n and the property that *all* functions in Ω are rationally expressible in terms of α, β. Then let

$$F(\alpha, \beta) = a_0 \alpha^n + a_1 \alpha^{n-1} + \cdots + a_{n-1}\alpha + a_n$$
$$= b_0 \beta^m + b_1 \beta^{m-1} + \cdots + b_{m-1}\beta + b_m = 0$$

be the irreducible equation between α, β, where $a_0, a_1, \ldots, a_n$ are polynomial functions of β and $b_0, b_1, \ldots, b_m$ are likewise polynomial functions of α.

Also let

$$\alpha = \frac{\mathfrak{A}_1}{\mathfrak{A}}, \quad \beta = \frac{\mathfrak{B}_1}{\mathfrak{B}},$$

where $\mathfrak{A}_1$ is relatively prime to $\mathfrak{A}$ and $\mathfrak{B}_1$ is relatively prime to $\mathfrak{B}$, so that $\mathfrak{A}, \mathfrak{A}_1$ are of order m, and $\mathfrak{B}, \mathfrak{B}_1$ are of order n. Now

$$F'(\alpha) = na_0 \alpha^{n-1} + (n-1)a_2 \alpha^{n-2} + \cdots + a_{n-1},$$
$$\alpha F'(\alpha) = -a_1 \alpha^{n-1} - 2a_2 \alpha^{n-2} - \cdots - na_n,$$

whence it follows that

$$F'(\alpha) = \frac{\mathfrak{K}}{\mathfrak{A}^{n-2}\mathfrak{B}^m}$$

and similarly

$$F'(\beta) = \frac{\mathfrak{L}}{\mathfrak{A}^n \mathfrak{B}^{m-2}}.$$

We shall now prove that the polygon $\mathfrak{K}$ is divisible by $\mathfrak{Z}_\beta$, and $\mathfrak{L}$ by $\mathfrak{Z}_\alpha$.

For $\mathfrak{K}$ this is easy to see under the hypothesis that at all points of $\mathfrak{Z}_\beta$ the function β has a finite value and a_0 has a nonzero value. Because

$$\alpha' = a_0 \alpha$$

is an integral algebraic function of β, and when one sets

$$f(\alpha') = a^{n-1}F(\alpha, \beta)$$

one has

$$f'(\alpha') = a_0^{n-2}F'(\alpha).$$

Now since $\mathfrak{o}_\beta f'(\alpha')$ is divisible by the ramification ideal in β generated by $\mathfrak{Z}_\beta$, by §11, 5, the correctness of the assertion follows. An analogous result holds for $F'(\beta)$.

If one now subjects α, β to arbitrary linear transformations:

$$\alpha = \frac{c + d\alpha'}{a + b\alpha'}, \quad \beta = \frac{c' + d'\beta'}{a' + b'\beta'},$$

$$(a + b\alpha')(d - b\alpha) = ad - bc,$$

$$(a' + b'\beta')(d' - b'\beta) = a'd' - b'c',$$

then

$$3_\alpha = 3_{\alpha'}, \quad 3_\beta = 3_{\beta'}$$

by §16, 2, and the irreducible equation between α', β' reads:

$$F_1(\alpha', \beta') = (a + b\alpha')^n (a' + b'\beta')^m F(\alpha, \beta) = 0.$$

But it is possible under all circumstances to choose the constants $a, b, c, d; a', b', c', d'$ so that the above hypotheses are satisfied for both α' and β'.

If one puts the coefficients a_0', b_0' of α'^n, β'^m in $F_1(\alpha', \beta')$ in the form

$$a_0' = (a' + b'\beta')^m (a_0 d^n + a_1 d^{n-1} b + \cdots + a_n b^n)$$

$$= \left(\frac{a'd' - b'c'}{d' - b'\beta} \right)^m (a_0 d^n + a_1 d^{n-1} b + \cdots + a_n b^n),$$

$$b_0' = (a + b\alpha')^n (b_0 d'^m + b_1 d'^{m-1} b' + \cdots + b_m b'^m)$$

$$= \left(\frac{ad - bc}{d - b\alpha} \right)^n (b_0 d'^m + b_1 d'^{m-1} b' + \cdots + b_m b'^m),$$

then one easily sees that the functions $a_0', d' - b'\beta$ can vanish at a point of 3_β, and $b_0', d - b\alpha$ can vanish at a point of 3_α, for only a finite number of values of the ratios $d : b, d' : b'$.

If we now set

$$\alpha' = \frac{\mathfrak{A}_1'}{\mathfrak{A}'}, \quad \beta' = \frac{\mathfrak{B}_1'}{\mathfrak{B}'},$$

then it follows (§19, 1) that

$$d - b\alpha = \frac{\mathfrak{A}_2}{\mathfrak{A}}, \quad a + b\alpha' = \frac{\mathfrak{A}_2'}{\mathfrak{A}'},$$

so:

$$\mathfrak{A}_2 \mathfrak{A}_2' = \mathfrak{A}\mathfrak{A}'.$$

However, if, as assumed, b is nonzero, then $\mathfrak{A}_2$ is relatively prime to $\mathfrak{A}$, because at a point of $\mathfrak{A}$ the order number of $d - b\alpha$ is the same as that of α (§15, 5) and hence

$$\mathfrak{A}_2 = \mathfrak{A}', \quad \mathfrak{A}_2' = \mathfrak{A},$$

so:

$$a + b\alpha' = \frac{\mathfrak{A}}{\mathfrak{A}'},$$

and similarly

$$a' + b'\beta' = \frac{\mathfrak{B}}{\mathfrak{B}'}.$$

But now, since $F(\alpha, \beta) = 0$, we have

$$F_1'(\alpha') = (ad - bc)(a + b\alpha')^{n-2}(a' + b'\beta')^m F'(\alpha),$$

and thus when

$$F_1'(\alpha') = \frac{\mathfrak{R}3_\beta}{\mathfrak{A}'^{n-2}\mathfrak{B}'^m},$$

as assumed, it follows that

$$F'(\alpha) = \frac{\mathfrak{R}\mathfrak{Z}_\beta}{\mathfrak{A}^{n-2}\mathfrak{B}^m}$$

and similarly

$$F'(\beta) = \frac{\mathfrak{R}\mathfrak{Z}_\alpha}{\mathfrak{A}^n\mathfrak{B}^{m-2}}.$$

The same polygon $\mathfrak{R}$ necessarily occurs in the numerators of both expressions because

$$\frac{d\alpha}{d\beta} = -\frac{F'(\beta)}{F'(\alpha)} = \frac{\mathfrak{B}^2\mathfrak{Z}_\alpha}{\mathfrak{A}^2\mathfrak{Z}_\beta}.$$

Now the order of the polygon $\mathfrak{A}^{n-2}\mathfrak{B}^m$ is

$$m(n-2) + mn = 2m(n-1),$$

thus the order of $\mathfrak{R}$,

$$2r = 2m(n-1) - w_\beta,$$

is always even, and this implies

(3)
$$p = \frac{1}{2}w_\beta - n + 1 = (n-1)(m-1) - r.$$

The polygon $\mathfrak{R}$ is called the *polygon of the double points* in (α, β).

§25. The differentials in Ω

SUMMARY AND COMMENTS

In this section Dedekind and Weber extend the concept of differential from the differentials $d\alpha$ (for $\alpha \in \Omega$), introduced in §23, to more general differentials of the form ωdz. The latter are defined so as to be independent of the variable z and are called *differentials in* Ω. The differentials $d\alpha$ from §23 are now called *proper* differentials, while the new ones ωdz (if not of the form $d\alpha$) are called *improper* or *Abelian* differentials.

If z, z_1 are any two variables in Ω, with orders n, n_1 and ramification numbers w, w_1, and if $\mathfrak{Z}, \mathfrak{Z}_1$ are the ramification polygons and $\mathfrak{U}, \mathfrak{U}_1$ are the lower polygons[48] of z, z_1 then (§23)

(1)
$$\frac{dz}{dz_1} = \frac{\mathfrak{Z}\mathfrak{U}_1^2}{\mathfrak{Z}_1\mathfrak{U}^2}.$$

Each function ω in Ω may be put in the form

(2)
$$\omega = \frac{\mathfrak{U}^2\mathfrak{A}}{\mathfrak{Z}\mathfrak{B}},$$

where $\mathfrak{A}, \mathfrak{B}$ are polygons whose orders a, b satisfy the condition

$$2n + a = w + b$$

or (§24)

(3)
$$a = b + 2p - 2.$$

---

[48]The letter $\mathfrak{U}$ (Fraktur U) stands for the German word "unter," meaning lower or under. (Translator's note.)

Now when one defines a function ω_1 by the equation

$$\omega\, dz = \omega_1\, dz_1,$$

then ω_1 acquires the description

$$\omega_1 = \frac{\mathfrak{U}_1^2}{\mathfrak{Z}_1 \mathfrak{B}}$$

from (1). In what follows we shall call expressions such as

$$\omega\, dz = \omega_1\, dz_1$$

differentials in Ω, and denote them by symbols such as $d\tilde{\omega}$. Such a differential is thereby defined *invariantly*, i.e., independently of the choice of variable z, and is completely determined by the two polygons $\mathfrak{U}, \mathfrak{B}$.

Without danger of misunderstanding we can use the symbolic notation

$$d\tilde{\omega} = \frac{\mathfrak{U}}{\mathfrak{B}},$$

and hence also, for example,

$$dz = \frac{\mathfrak{Z}}{\mathfrak{U}^2}.$$

This notation for a differential as a quotient of polygons is distinguished from the similar notation for functions in Ω (§17) by the fact that in the latter the numerator and denominator are of the same order, while in differentials the order of the numerator exceeds that of the denominator by $2p - 2$. As with the notation of §17, any common divisor of $\mathfrak{U}$ and $\mathfrak{B}$ can be cancelled out. If $\mathfrak{U}$ and $\mathfrak{B}$ are relatively prime, then $\mathfrak{U}$ is called the upper polygon, and $\mathfrak{B}$ the lower polygon, of the differential $d\tilde{\omega}$.

The general concept of differentials in Ω presented here includes the differentials of functions in the field Ω, defined in §23, 4. We call the latter *proper differentials*, while the others, which cannot be represented as differentials of functions existing in Ω will be called *improper* or *Abelian differentials*.

Functions of the form (2), which in our newly adopted notation may be denoted by $\frac{d\tilde{\omega}}{dz}$, will be called *differential quotients with respect to* z, and we similarly distinguish between *proper* and *improper* differential quotients, according as $d\tilde{\omega}$ is a proper or improper differential.[49]

There now arises the problem of determining the extent of the differential concept, i.e., of finding all polygons $\mathfrak{U}, \mathfrak{B}$ that can be upper and lower polygons of a differential. We begin with the following general remarks:

A necessary and sufficient condition for $\frac{\mathfrak{U}}{\mathfrak{B}}$ to be a differential is that

$$\frac{\mathfrak{U}^2 \mathfrak{U}}{\mathfrak{Z} \mathfrak{B}}$$

be a function in Ω for an arbitrary variable z, in other words, that $\mathfrak{U}^2\mathfrak{U}$ be equivalent to $\mathfrak{Z}\mathfrak{B}$. This relationship is preserved, however, when $\mathfrak{U}, \mathfrak{B}$ themselves are replaced by equivalent polygons $\mathfrak{U}', \mathfrak{B}'$. If we hold $\mathfrak{B}$ fixed, and if $\frac{\mathfrak{U}}{\mathfrak{B}}$ is a differential, then

$$\frac{\mathfrak{U}'}{\mathfrak{B}}, \quad \frac{\mathfrak{U}''}{\mathfrak{B}}, \quad \cdots$$

---

[49]The quotient of any two proper or improper differentials $\frac{d\tilde{\omega}}{d\tilde{\omega}'}$ always has meaning as a definite function in Ω. In what follows we confine ourselves to the consideration of quotients in which at least the denominator is a proper differential.

represent differentials if and only if the polygons $\mathfrak{A}, \mathfrak{A}', \mathfrak{A}'', \ldots$ all belong to the same class A. If the polygons $\mathfrak{A}_1, \mathfrak{A}_2, \mathfrak{A}_3, \ldots$ form a basis of A, so that

$$A = (\mathfrak{A}_1, \mathfrak{A}_2, \mathfrak{A}_3, \ldots),$$

then the corresponding differential quotients $\frac{d\tilde{\omega}_1}{dz}, \frac{d\tilde{\omega}_2}{dz}, \frac{d\tilde{\omega}_3}{dz}, \ldots$ with respect to any variable z form a basis for a finite-dimensional vector space of functions, and we likewise say that $d\tilde{\omega}_1, d\tilde{\omega}_2, d\tilde{\omega}_3, \ldots$ is a basis for a *vector space of differentials*

$$(d\tilde{\omega}_1, d\tilde{\omega}_2, d\tilde{\omega}_3, \ldots).$$

This means that each differential $d\tilde{\omega}$ whose lower polygon is $\mathfrak{B}$ or a divisor of $\mathfrak{B}$ may be expressed in the form, with constant coefficients $c_1, c_2, c_3, \ldots$,

$$d\tilde{\omega} = c_1 d\tilde{\omega}_1 + c_2 d\tilde{\omega}_2 + c_3 d\tilde{\omega}_3 + \cdots.$$

§26. Differentials of the first kind

SUMMARY AND COMMENTS

Differentials are classified into three *kinds*, of which the first is treated here. This classification goes back to the classification of elliptic integrals into three kinds by Legendre in 1793. The differentials of the first kind are those without poles. Those of the second and third kind are defined in §31.

In this section, Dedekind and Weber prove Abel's theorem for differentials of the first kind. Differentials of the second and third kind are treated in §31. However, the version of Abel's theorem here is hardly recognizable as such, since it does not even mention the genus p. It does have a consequence (the "fundamental theorem") involving p, but a theorem resembling Abel's statement does not appear until §33, where Abel's name does not even occur.

As mentioned in §10, the proof of Abel's theorem hinges on the concept of *complementary basis* introduced there. It is now joined by the related concept of *supplementary polygon*, defined to be polygons whose product is *complete*, that is, the upper polygon of a differential. Likewise, along with differentials of the first kind, there are polygons of the first kind. There are also polygons of the second kind. But in the case of polygons, this simply means "not of the first kind."

Recalling the concept of a polygon *class* from §18, Dedekind and Weber introduce the class W of complete polygons of the first kind and prove that it has dimension p. In §27 they name it the *principal* class. W happens to be none other than the special class singled out by Roch in his contribution to the Riemann-Roch theorem. It is known today as the *canonical class*.

We first consider the simplest differentials in Ω, namely those for which the lower polygon is the null-gon $\mathfrak{O}$. Such differentials (whose existence is admittedly not proved yet) are called *differentials of the first kind*. The upper polygon $\mathfrak{W}$ of such a differential dw, whose order is $2p - 2$, is called the *fundamental polygon* of

dw and it is said to be a *complete polygon of the first kind*, while each divisor of such a polygon is simply called a *polygon of the first kind*. If $\mathfrak{W} = \mathfrak{A}\mathfrak{B}$ then $\mathfrak{A}, \mathfrak{B}$ are called *supplementary polygons* of each other. A polygon that is not a divisor of a complete polygon of the first kind, in particular any polygon with more than $2p - 2$ points, is called a *polygon of the second kind*.

1. By the above remarks, all complete polygons of the first kind form a polygon class W, whose dimension remains to be determined. If this dimension turns out to be > 0, then the existence of polygons of the first kind is proved at the same time. But the latter dimension is the same as the dimension of the vector space of differentials of the first kind, which is also the same as that of the vector space of *differential quotients of the first kind*, for an arbitrary variable z, when we take the differential quotients of the first kind with respect to z to be the functions

$$u = \frac{dw}{dz}.$$

By §25, (2), such a function u is expressible as

$$u = \frac{\mathfrak{A}^2 \mathfrak{W}}{3},$$

and one easily sees, by consideration of the order numbers at the different points, that such a differential quotient of the first kind is completely defined by the following two properties:

I. At each point $\mathfrak{P}$ where z has a finite value z_0,

$$(u(z - z_0))_0 = 0.$$

II. At a point $\mathfrak{P}$ where z is infinite,

$$(zu)_0 = 0.$$

If, as in §11, 4,

$$r = (z - c)(z - c_1)(z - c_2) \cdots$$

is the product of all the distinct linear factors of the discriminant $\Delta_z(\Omega)$ and $\mathfrak{r}$ is the product of all the distinct prime ideals dividing r, then condition I is equivalent to the fact that ru is a function in $\mathfrak{r}$, or that u is a function in the module $\mathfrak{e}$ complementary to $\mathfrak{o}$ (§11, 4(6)). Thus in order to obtain the totality of functions u one has to look among the functions in $\mathfrak{e}$ for those that satisfy condition II.

2. For this purpose we take a *normal basis* $\lambda_1, \lambda_2, \ldots, \lambda_n$ of $\mathfrak{o}$ (§22) and denote its complementary basis by $\mu_1, \mu_2, \ldots, \mu_n$, so that each function satisfying condition I, hence also each differential quotient of the first kind, is of the form

(1) $$u = y_1 \mu_1 + y_2 \mu_2 + \cdots + y_n \mu_n,$$

where $y_1, y_2, \ldots, y_n$ are *polynomial* functions of z. But it follows from the basic properties of the complementary basis that (§10, 3)

$$y_s = Tr(u\lambda_s); \qquad \frac{y_s}{z^{r_s - 1}} = Tr\left(uz \cdot \frac{\lambda_s}{z^{r_s}}\right).$$

Now since $\frac{\lambda_s}{z^{r_s}}$ is in $\mathfrak{o}'$, and hence finite for $z = \infty$, and since uz vanishes at each such point by II, it follows (§16, 5) that $\frac{y_s}{z^{r_s - 1}}$ must vanish for $z = \infty$, i.e., the degree of the polynomial function y cannot exceed $r_s - 2$.

Therefore, if $r_s < 2$, y must vanish, and hence

$$y_1 = 0; \quad Tr(u) = 0$$

under all circumstances (§22, 2) (*Abel's theorem for differentials of the first kind*) and, if $r_s \geq 2$:

(2) $$y_s = c_0 + c_1 z + c_2 z^2 + \cdots + c_{r_s - 2} z^{r_s - 2}.$$

It remains to show that these conditions are also sufficient, i.e., that each function of the form (1) in which y_s has the form (2) satisfies condition II, or, in other words, that when $r_s \geq 2$, $z^{r_s - 1}\mu_s$ vanishes at all points at which z is infinite. This follows immediately by consideration of the system $\mathfrak{o}'$ of integral algebraic functions of $z' = \frac{1}{z}$, for which the functions

$$\lambda_1' = \frac{\lambda_1}{z^{r_1}}, \quad \lambda_2' = \frac{\lambda_2}{z^{r_2}}, \quad \ldots, \quad \lambda_n' = \frac{\lambda_n}{z^{r_n}}$$

form a normal basis by §22, 3. By §10, 5 the basis complementary to this is

$$\mu_1' = z^{r_1}\mu_1, \quad \mu_2' = z^{r_2}\mu_2, \quad \ldots, \quad \mu_n' = z^{r_n}\mu_n,$$

and since

$$z'\mu_s' = 0 \quad \text{for} \quad z' = 0$$

(by property I, applied to z', μ'), it follows that

$$z^{r_s - 1}\mu_s = 0 \quad \text{for} \quad z = \infty.$$

Q.E.D.

But since the functions $z^h \mu_s$ are linearly independent (by the rational independence of the functions μ_s), §24, (2) yields the *fundamental theorem*:

The vector space of differentials of the first kind is of dimension

$$(r_2 - 1) + (r_3 - 1) + \cdots + (r_n - 1) = p,$$

and hence p is also the dimension of the class W of complete polygons of the first kind.

As a basis for the vector space of differential quotients of the first kind with respect to z one can choose the p functions $z^h \mu_s$ ($h \leq r_s - 2$), and the fundamental polygons $\mathfrak{W}_1, \mathfrak{W}_2, \ldots, \mathfrak{W}_p$ for the corresponding differentials dw form a basis of the class W.

3. For the sake of a later application we now consider a special type of differential quotient u' of the first kind, namely that in which condition II is replaced by the following condition:

III. At each point $\mathfrak{P}$ where z is infinite,

$$(z^k u')_0 = 0,$$

where k is a given positive integer.

The functions u' may be expressed as

$$u' = \frac{\mathfrak{A}^{k+1}\mathfrak{W}'}{3}$$

and they likewise form a vector space. The polygons $\mathfrak{W}'$ also form a class W', whose order is

$$w - n(k + 1) = 2p - 2 - n(k - 1).$$

However, the polygons $\mathfrak{W}'$ are *not* independent of the choice of variable z. The dimension of the class W' may be determined by the same method as for the class W. Since condition I is satisfied, the functions u' are likewise expressible in the form (1). However,

$$\frac{y_s}{z^{r_s-k}} = Tr\left(u' z^k \frac{\lambda_s}{z^{r_s}}\right)$$

must now vanish for $z = \infty$, and therefore the degree of the polynomial y_s cannot exceed the number $r_s - k - 1$. Thus y_s vanishes identically as soon as $r_s < k + 1$. Otherwise

(3) $$y_s = c_0 + c_1 z + \cdots + c_{r_s-k-1} z^{r_s-k-1}.$$

Conversely, if y_s has this form, then condition III is satisfied by the function

$$u' = \sum_s y_s \mu_s,$$

because

$$z^k(z^{r_s-k-1}\mu_s) = z^{r_s-1}\mu_s$$

has the value 0 for $z = \infty$, as proved in 2.

It follows that the dimension of the vector space of functions u', and hence also that of the class W', is

$$\sum_\iota (r_\iota - k),$$

where the sum is taken only over terms that have a positive value. If all $r_\iota - k \leq 0$ the desired functions do not exist at all.

§27. Polygon classes of the first and second kind

Summary and Comments

In this section, the concepts of "supplementary," "first kind," and "second kind" in §26 are extended from polygons to their equivalence classes. In particular, classes A and B are called supplementary if $AB = W$, where W is the canonical class introduced in §26 and called (by Dedekind and Weber) the *principal* class. The dimensions of supplementary classes are also investigated, with an eye to the proof of the Riemann-Roch theorem in the next section.

If $\mathfrak{A}$ is a polygon of the first kind then all polygons equivalent to $\mathfrak{A}$ are likewise of the first kind. Because when $\mathfrak{A}$ and $\mathfrak{B}$ are supplementary polygons and

$$\mathfrak{A}\mathfrak{B} = \mathfrak{W}$$

then

$$AB = W,$$

where A, B are the classes of $\mathfrak{A}$ and $\mathfrak{B}$, and when $\mathfrak{A}'$ is equivalent to $\mathfrak{A}$, $\mathfrak{A}'\mathfrak{B} = \mathfrak{W}'$ is also equivalent to $\mathfrak{W}$ (§18, 5).

We therefore call such classes, which contain polygons of the first kind, *polygon classes of the first kind*. The rest are called *polygon classes of the second kind*. The class W of complete polygons of the first kind is called the *principal class*, and two classes A, B that satisfy the condition

$$AB = W$$

are called *supplementary classes*.

If

$$\eta = \frac{\mathfrak{A}'}{\mathfrak{A}}$$

is a function in Ω and $\mathfrak{A}'$ is relatively prime to $\mathfrak{A}$, so that the class A of $\mathfrak{A}$ is proper, we call η a function of the *first* or *second* kind according as the class A is of the first or second kind.

If A is an arbitrary class of the first kind and q is the number of independent polygons $\mathfrak{W}$ divisible by a polygon $\mathfrak{A}$ in the class A then, by §21, 2,

$$q = (A, W) = (O, B),$$

i.e., it equals the dimension of the supplementary class B of A. Similarly, (B, W) equals the dimension of the class A. If A is a class of the second kind then $(A, W) = 0$. Since p is the dimension of W, then by §20, 2, 3 each class of order $\leq p - 1$ is of the first kind, and in particular there are classes A of order $p - k$ such that $(A, W) = (O, B) = k$. It follows from the same theorems that there are classes of order p that are of the second kind.

§28. The Riemann-Roch theorem for proper classes

SUMMARY AND COMMENTS

In this section, the most substantial part of the proof of the Riemann-Roch theorem—for proper classes of the first kind—is carried out. Dedekind and Weber first express the theorem in terms of supplementary classes, then (in subsection 5) in terms of the canonical class W, as is more usual today. They also prove that W is always proper.

The Riemann-Roch theorem, which in its usual formulation gives the number of arbitrary constants in a function that is infinite at a certain number of given points, in our presentation expresses a relation between the dimension and the order of a class, or between a class and its supplementary class. Confining ourselves initially to *proper* classes, we preface the derivation of this fundamental relation with the following remarks.

1. In a proper class A it is always possible, by §19, 2, to choose two relatively prime polygons $\mathfrak{A}, \mathfrak{A}'$ (one of which can be an arbitrary member of the class). Thus if one sets

$$z = \frac{\mathfrak{A}'}{\mathfrak{A}},$$

and if $\mathfrak{A}''$ is an arbitrary third polygon in the class A, then

$$\omega = \frac{\mathfrak{A}''}{\mathfrak{A}}, \quad \frac{\omega}{z} = \frac{\mathfrak{A}''}{\mathfrak{A}'}.$$

Hence, by §17, ω is an *integral algebraic* function of z and $\frac{\omega}{z}$ is an integral algebraic function of $\frac{1}{z}$. It then follows from §22 that the exponent of ω is ≤ 1.

Conversely, if ω is an integral algebraic function of z with exponent < 1 then it has the form

$$\omega = \frac{\mathfrak{A}''}{\mathfrak{A}},$$

where $\mathfrak{A}''$ is a polygon in the class A. Because if

$$\omega = \frac{\mathfrak{A}_1''}{\mathfrak{A}_1}, \qquad \frac{\omega}{z} = \frac{\mathfrak{A}_1''\mathfrak{A}}{\mathfrak{A}_1\mathfrak{A}'}$$

and $\mathfrak{A}_1''$ is taken relatively prime to $\mathfrak{A}_1$ then, since ω is an integral algebraic function of z, it follows first that $\mathfrak{A}_1$ cannot contain points that are not also in $\mathfrak{A}$. But $\mathfrak{A}_1$ also cannot contain points more often than $\mathfrak{A}$ does, otherwise $\frac{\omega}{z}$ would be infinite at such a point (which cannot appear in $\mathfrak{A}'$), and hence would not be an integral algebraic function of $\frac{1}{z}$. Therefore $\mathfrak{A}$ is divisible by $\mathfrak{A}_1$, and ω can be put in the form $\frac{\mathfrak{A}''}{\mathfrak{A}}$.

2. Thus in order to obtain the totality of polygons in the class A it suffices to seek the integral algebraic functions of z with exponent ≤ 1.

If n is the order of the class A, and hence also the order of the variable z, and if $\lambda_1, \lambda_2, \ldots, \lambda_n$ form a normal basis of $\mathfrak{o}$ with exponents $r_1, r_2, \ldots, r_n$, amongst which r_s is the last ≤ 1, then each function ω with exponent ≤ 1 can be expressed in the form

$$\omega = c_1\lambda_1 + c_2\lambda_2 + \cdots + c_s\lambda_s + z\omega_1,$$

by §22, 2. But since the exponent of $z\omega_1$ cannot be larger than 1, ω_1 must be a constant and therefore

$$\omega = c_1\lambda_1 + c_2\lambda_2 + \cdots + c_s\lambda_s + c_{s+1}z.$$

Conversely, each function of this form satisfies the stated condition. Thus $s + 1$ is the dimension of the class A, which therefore, in agreement with §21, 1, is always $\leq n + 1$. The upper bound $n + 1$ can only be attained in the case $p = 0$, but it is actually attained, because in this case $r_2, r_3, \ldots, r_n = 1$. It follows that, when $p = 0$, *only a single point* $\mathfrak{P}$ can belong to a proper class.

3. If one of the exponents $r_{s+1}, r_{s+2}, \ldots, r_n$ is greater than 2 then certainly $r_n > 2$ also. Hence, by §26, 2, the following are differential quotients of the first kind with respect to z:

$$\mu_n = \frac{\mathfrak{A}^2\mathfrak{W}}{\mathfrak{Z}}, \qquad \mu_n z = \frac{\mathfrak{A}^2\mathfrak{W}_1}{\mathfrak{Z}} = \frac{\mathfrak{A}\mathfrak{A}'\mathfrak{W}}{\mathfrak{Z}},$$

where $\mathfrak{Z}$ is the ramification polygon for z. Thus

$$\mathfrak{A}\mathfrak{W}_1 = \mathfrak{A}'\mathfrak{W}$$

or, since $\mathfrak{A}, \mathfrak{A}'$ are relatively prime,

$$\mathfrak{W} = \mathfrak{A}\mathfrak{B}, \qquad \mathfrak{W}_1 = \mathfrak{A}'\mathfrak{B}.$$

I.e., the class A is of the *first* kind (z is a variable of the first kind). Thus if we first assume that A is a class of the *second* kind then it follows that

$$r_{s+1} = 2, \qquad r_{s+2} = 2, \qquad \ldots, \qquad r_n = 2$$

and

$$p = (r_2 - 1) + \cdots + (r_s - 1) + (r_{s+1} - 1) + \cdots + (r_n - 1) = n - s.$$

The dimension $s + 1$ of the class A is therefore

$$(O, A) = n - p + 1.$$

4. If we next assume that A is of the *first* kind and

$$q = (A, W)$$

as in §27, then there exist q linearly independent complete polygons of the first kind divisible by $\mathfrak{A}$, and the corresponding differential quotients of the first kind with respect to z, which are also q in number but no longer linearly independent, have the form

$$v = \frac{\mathfrak{A}^3 \mathfrak{B}}{3},$$

where $\mathfrak{B}$ is a polygon of $2p - 2 - n$ points. The class B of $\mathfrak{B}$ is the supplementary class of A, and therefore its dimension equals q (§27).

These functions v have the property, however, that at the vertices of $\mathfrak{A}$, i.e., where $z = \infty$, not only does zv vanish but so too does

$$z^2 v = \frac{\mathfrak{A}\mathfrak{A}'^2 \mathfrak{B}}{3}.$$

It follows that they are completely determined by the property of being differential quotients of the first kind. Because if

$$v = \frac{\mathfrak{A}^2 \mathfrak{W}}{3}, \quad vz^2 = \frac{\mathfrak{A}'^2 \mathfrak{W}}{3},$$

then if $z^2 v$ is to vanish at all points of $\mathfrak{A}$, $\mathfrak{W}$ must be divisible by $\mathfrak{A}$, since $\mathfrak{A}'$ is assumed to be relatively prime to $\mathfrak{A}$. Hence by §26, 3

$$q = (r_{s+1} - 2) + (r_{s+2} - 2) + \cdots + (r_n - 2),$$

moreover

$$p = (r_{s+1} - 1) + (r_{s+2} - 1) + \cdots + (r_n - 1),$$

hence:

$$p - q = n - s, \quad s = n - p + q.$$

This contains the *Riemann-Roch theorem*, which we can express as follows in the present case, in view of §27: *if A, B are supplementary classes of the first kind, at least one of which is proper, and if a, b are their orders, so that*

$$a + b = 2p - 2,$$

then

$$(O, A) - \frac{1}{2} a = (O, B) - \frac{1}{2} b.$$

5. When the case $(A, W) = 0$ is not excluded, the Riemann-Roch theorem comprehending both cases is:

If A is a proper class of order n then its dimension is

$$(O, A) = n - p + 1 + (A, W).$$

Since the dimension of a proper class must be at least 2 (when it does not consist of a single null-gon), it follows that, when $(A, W) = 0$,

$$n \geq p + 1,$$

whence we have the theorem due to Riemann:

Each function of order $< p$ is a function of the first kind.

6. With the help of this theorem it is easy to prove that *the principal class W of complete polygons of the first kind is always proper.*

Namely, if $\mathfrak{M}$ is the divisor of W then by §19, 2 it is possible to find a polygon $\mathfrak{A}\mathfrak{M}$ in W such that $\mathfrak{A}$ is relatively prime to $\mathfrak{M}$. The class A of $\mathfrak{A}$ is proper (§21, 3) and at the same time $\mathfrak{A}\mathfrak{M}$ is the only polygon in W divisible by $\mathfrak{A}$ (since each polygon in W has the divisor $\mathfrak{M}$). Thus

$$(A, W) = 1.$$

Now p is the dimension of W, and hence also that of A. So, by the Riemann-Roch theorem, the order of A equals $2p - 2$, i.e., it is as large as that of W. Consequently $\mathfrak{M} = \mathfrak{O}$.

§29. The Riemann-Roch theorem for improper classes of the first kind

SUMMARY AND COMMENTS

The proof of the Riemann-Roch theorem is now extended to improper classes of the first kind, by reducing it to the case of proper classes.

If A is a class of the first kind with divisor $\mathfrak{M}$, and if

$$A = \mathfrak{M}A',$$

then A' is a proper class of the first kind. Let B be the supplementary class of A; B' that of A'; a, b the orders of A, B; and m the order of M. One obtains the whole class B when one cancels the factor $\mathfrak{M}$ in all the polygons in B' divisible by $\mathfrak{M}$. Because if

$$\mathfrak{A}\mathfrak{B} = \mathfrak{A}'\mathfrak{M}\mathfrak{B} = \mathfrak{W},$$

then $\mathfrak{M}\mathfrak{B}$ belongs to the class B' and conversely, if

$$\mathfrak{A}'\mathfrak{B}' = \mathfrak{A}'\mathfrak{M}\mathfrak{B} = \mathfrak{W}$$

then $\mathfrak{B}$ belongs to the class B.

But by §21, 2 this implies

$$(O, B) \geq (O, B') - m.$$

Now A' is a proper class of the same dimension as A and of order $a - m$, hence (§28, 5)

$$(O, A) = (O, A') = a - m - p + 1 + (A', W),$$

or

$$(O, A) = (O, B') - m + a - p + 1.$$

Therefore

$$(O, A) \leq (O, B) + a - p + 1 = (O, B) + \frac{1}{2}(a - b),$$

hence

$$(O, A) - \frac{1}{2}a \leq (O, B) - \frac{1}{2}b.$$

But since the classes A, B can be interchanged, we similarly get

$$(O, B) - \frac{1}{2}b \leq (O, A) - \frac{1}{2}a,$$

i.e.,

$$(O, A) - \frac{1}{2}a = (O, B) - \frac{1}{2}b,$$

so we have proved the Riemann-Roch theorem in the same form as in §28, 4 for general polygon classes of the first kind.[50] [51]

§30. Improper classes of the second kind

SUMMARY AND COMMENTS

The proof of the Riemann-Roch theorem is now completed, by studying subcases of the remaining case.

We now seek the conditions under which a polygon class A of the second kind and order n can actually be improper, which will yield the general validity of the Riemann-Roch theorem.

1. Each class A can always be converted to a proper class AN by multiplication by another class N of order ν. Because if $\mathfrak{A}$ is an arbitrary polygon in A one chooses a variable z that remains finite at all the points of $\mathfrak{A}$ (§15, 6). Then, if η is an arbitrary function in the ideal with respect to z generated by $\mathfrak{A}$, the upper polygon of η is divisible by $\mathfrak{A}$, hence of the form $\mathfrak{A}\mathfrak{N}$, and the class of $\mathfrak{A}\mathfrak{N}$ is proper.

2. The dimension of the proper class AN of the second kind is

$$(O, AN) = n + \nu - p + 1$$

by §28, 3, and then it follows by §21, 2 that

$$(O, A) \geq n - p + 1.$$

Now if the divisor $\mathfrak{M}$ of the class A is of order m, and if

$$A = \mathfrak{M}A',$$

then A' is a proper class of the same dimension as A, and hence (§28, 5)

$$(O, A) = (O, A') = n - m - p + 1 + (A', W),$$

so

$$(A', W) \geq m,$$

i.e., A' must certainly be of the first kind when A is an improper class. Thus if B' is the supplementary class of A' we also have

$$(O, B') \geq m.$$

But if $(O, B') > m$ it is possible, by §20, 2, to find a polygon $\mathfrak{M}\mathfrak{B}$ in B' divisible by $\mathfrak{M}$, and

$$\mathfrak{A}'\mathfrak{M}\mathfrak{B} = \mathfrak{A}\mathfrak{B} = \mathfrak{W}$$

---

[50]In the terminology of Christoffel (*On the canonical form of Riemann's integral of the first kind*, Annali di Mathematica pura et applicata, Serie II, Tomo IX)

$$(A, W) + a - p = (O, B) + a - p = (O, A) - 1$$

is the "excess" and

$$(A, W) - 1 = (O, B) - 1$$

is the "defect" of the point system $\mathfrak{A}$.

[51]Here Dedekind and Weber use Christoffel's term "point system" for what they elsewhere call a "polygon." (Translator's note.)

so A is of the first kind, contrary to hypothesis. Thus

$$(A', W) = m$$

and hence

$$(O, A) = n - p + 1,$$

so in this case we again get the Riemann-Roch theorem, precisely in the form of §28, 3.

3. If the class A contains only a single isolated polygon, then $(O, A) = n-p+1$, hence $n = p$. I.e., an isolated polygon of the second kind always has order p. Conversely, it follows from 2 that each polygon of the second kind and order p is isolated.

4. Retaining the notation of 2, $(O, B') = m$, and hence by the frequently used theorem (§20, 2) there is a polygon in B' divisible by any $(m-1)$-gon. Thus if one sets

$$\mathfrak{M} = \mathfrak{P}\mathfrak{M}',$$

detaching an arbitrary point $\mathfrak{P}$ from $\mathfrak{M}$, then there is a polygon $\mathfrak{M}'\mathfrak{B}$ in B and hence

$$\mathfrak{A}'\mathfrak{M}'\mathfrak{B} = \mathfrak{W}.$$

The polygon $\mathfrak{A}'\mathfrak{M}' = \mathfrak{A}''$ and its class A'' are therefore of the *first kind* and A has the form

$$A = PA'',$$

where P denotes the class of $\mathfrak{P}$. At the same time we must have $(A'', W) = (O, B'') = 1$, i.e., the supplementary class B'' of A'' contains only a single isolated polygon $\mathfrak{B}''$, since otherwise there would be a polygon in B'' divisible by $\mathfrak{P}$, and hence A would also be of the first kind, contrary to hypothesis.

5. Conversely, if A'' is a class of the first kind for which $(A'', W) = 1$, so that the supplementary class B'' of A'' consists of just an isolated polygon $\mathfrak{B}''$, and if $\mathfrak{P}$ is a point not appearing in $\mathfrak{B}''$, and its class is P, then $A = PA''$ is an improper class of the second kind and order n, divisible by $\mathfrak{P}$.

The fact that A is of the second kind comes from the assumption that $\mathfrak{P}$ does not divide $\mathfrak{B}''$. The dimension of A is therefore

$$(O, A) = n - p + 1$$

by 2, where n is the order of A. On the other hand, the dimension of the class A'' is

$$(O, A'') = n - p + (A'', W) = n - p + 1$$

by §§28 and 29, thus A and A'' are of the same dimension. But all the polygons in the class A'' become polygons in the class A under multiplication by $\mathfrak{P}$, and because of the equality of dimensions this completely exhausts the latter class. Thus all the polygons in the class A contain the factor $\mathfrak{P}$, which is therefore also in the divisor of A.

6. In the special case where the genus p of the field Ω is 0, polygons and classes of the first kind do not occur at all. Thus in this case there are also no improper classes. The dimension of each class is 1 greater than its order. In particular, each point $\mathfrak{P}$ belongs to a proper class of dimension 2, and hence in this case there are functions z in Ω of first order. In terms of one such function any other function in

the field is expressible *rationally*, because the irreducible equation relating z and the other variable is of first degree in the latter (§15, 7).

§31. Differentials of the second and third kinds

SUMMARY AND COMMENTS

The different kinds of differentials, the first of which was introduced in §26, are revisited in the light of the Riemann-Roch theorem. One now sees where differentials of the second and third kinds fit in.

1. Using the notation introduced in §25, suppose

$$d\tilde{\omega} = \frac{\mathfrak{A}}{\mathfrak{B}}$$

is an arbitrary differential in Ω. Thus

$$a = b + 2p - 2,$$

where a, b are the orders of $\mathfrak{A}, \mathfrak{B}$, and if $\mathfrak{A}, \mathfrak{B}$ are relatively prime and $\mathfrak{U}, \mathfrak{Z}$ denote the lower polygon and ramification polygon for an arbitrary variable z, then $\mathfrak{U}^2\mathfrak{A}$ must be equivalent to $\mathfrak{Z}\mathfrak{B}$ (§25). Thus if U, Z, A, B denote the classes of the polygons $\mathfrak{U}, \mathfrak{Z}, \mathfrak{A}, \mathfrak{B}$ we have

$$U^2 A = ZB.$$

On the other hand, when W is the principal class of the first kind,

$$U^2 W = Z,$$

which yields the relation

$$A = BW.$$

Conversely, if $\mathfrak{A}$ is an arbitrary polygon in the class BW then this relation implies the equivalence of $\mathfrak{U}^2\mathfrak{A}$ with $\mathfrak{Z}\mathfrak{B}$, and hence the existence of a differential denoted by $\frac{\mathfrak{A}}{\mathfrak{B}}$. It follows that $\mathfrak{B}$ is the lower polygon of a differential $d\tilde{\omega}$ if and only if BW contains a polygon relatively prime to $\mathfrak{B}$, i.e., when the divisor of the class BW is relatively prime to $\mathfrak{B}$. The dimension of the class BW is then also the dimension of the vector space of differentials $d\tilde{\omega}$ associated with the lower polygon $\mathfrak{B}$ (§25). Since $(W, W) = 1$, the theorems §30, 4, 5 then yield the following result.

a) If $\mathfrak{B}$ consists of a single point (i.e., if $b = 1$) then the class BW is improper with divisor $\mathfrak{B}$. *Thus the order b of the lower polygon of a differential $d\tilde{\omega}$ cannot be one.*

b) If $b \geq 2$ then BW is always a proper class of the second kind and hence its dimension is

$$b + p - 1.$$

The lower polygon of a differential can therefore be any polygon of more than one point, and $b + p - 1$ linearly independent differentials exist among those associated with a lower polygon of order b.

2. Under the hypothesis that $b \geq 2$, we now seek a basis for the class A such that each element $\mathfrak{A}_r$ of this basis yields a differential $d\tilde{\omega}_r$ of the simplest possible nature, namely, one whose lower polygon is a power of a single point or the product of only two different points.

Supposing that such a basis

(1) $$\mathfrak{A}_1, \mathfrak{A}_2, \mathfrak{A}_3, \ldots, \mathfrak{A}_{b+p-1}$$

has already been found for the class BW, we build a similar basis of dimension $b + p$ for the class BPW, namely

(2) $$\mathfrak{P}\mathfrak{A}_1, \mathfrak{P}\mathfrak{A}_2, \ldots, \mathfrak{P}\mathfrak{A}_{b+p-1}, \mathfrak{A}',$$

where P denotes the class of an arbitrary point $\mathfrak{P}$. The first $b + p - 1$ of these polygons really belong to the class BPW and are independent, because the polygons (1) are. Also, the differentials constructed from them,

$$d\tilde{\omega}_r = \frac{\mathfrak{P}\mathfrak{A}_r}{\mathfrak{P}\mathfrak{B}} = \frac{\mathfrak{A}_r}{\mathfrak{B}}$$

are identical with those constructed from (1). Thus it remains only to construct $\mathfrak{A}'$, for which we have to distinguish two cases.

a) If $\mathfrak{P}$ divides $\mathfrak{B}$ and $\mathfrak{B} = \mathfrak{M}\mathfrak{P}^m$, where $\mathfrak{M}$ is not divisible by $\mathfrak{P}$, then $P^{m+1}W$ is a proper class (because $m + 1 \geq 2$, §30, 4) in which there is consequently a polygon $\mathfrak{N}$ not divisible by $\mathfrak{P}$. If one now sets $\mathfrak{A}' = \mathfrak{M}\mathfrak{N}$ then $\mathfrak{A}'$ belongs to the class BPW and is not divisible by $\mathfrak{P}$. Hence it is also not in the vector space $(\mathfrak{P}\mathfrak{A}_1, \mathfrak{P}\mathfrak{A}_2, \ldots, \mathfrak{P}\mathfrak{A}_{b+p-1})$ whose divisor is $\mathfrak{P}$. It follows that the polygons (2) are independent of each other, and since there are $b + p$ of them, they form a basis for the class BPW. The differential constructed from $\mathfrak{A}'$,

$$d\tilde{\omega}' = \frac{\mathfrak{A}'}{\mathfrak{P}\mathfrak{B}} = \frac{\mathfrak{N}}{\mathfrak{P}^{m+1}},$$

has the required form, since its lower polygon is a power of a single point.

b) If $\mathfrak{P}$ does not divide $\mathfrak{B}$ one chooses once and for all a point $\mathfrak{P}_1$ dividing $\mathfrak{B}$ and sets $\mathfrak{B} = \mathfrak{M}\mathfrak{P}_1$ (whether or not $\mathfrak{M}$ is divisible by $\mathfrak{P}_1$). If one then chooses a polygon $\mathfrak{N}$ in the proper class PP_1W not divisible by $\mathfrak{P}$ and $\mathfrak{P}_1$ then $\mathfrak{A}' = \mathfrak{M}\mathfrak{N}$ again belongs to the class PBW, and since $\mathfrak{A}'$ is not divisible by $\mathfrak{P}$ it follows as above that the polygons (2) form a basis for BPW. At the same time

$$d\tilde{\omega}' = \frac{\mathfrak{A}'}{\mathfrak{P}\mathfrak{B}} = \frac{\mathfrak{N}}{\mathfrak{P}\mathfrak{P}_1}$$

hence it is of the required form.

It now only remains to describe the beginning of this operation. If $b = 0$, so that $\mathfrak{B} = \mathfrak{O}$, then

$$BW = W = (\mathfrak{W}_1, \mathfrak{W}_2, \ldots, \mathfrak{W}_p)$$

(the principal class of the first kind).

If $b = 2$ then one chooses, from the proper class BW, a polygon $\mathfrak{N}$ relatively prime to $\mathfrak{B}$. Then

$$BW = (\mathfrak{B}\mathfrak{W}_1, \mathfrak{B}\mathfrak{W}_2, \ldots, \mathfrak{B}\mathfrak{W}_p, \mathfrak{N}).$$

If one proceeds from this basis, in the way described above, to determine a basis (1) corresponding to an arbitrarily given polygon

$$\mathfrak{B} = \mathfrak{P}_1^{m_1} \mathfrak{P}_2^{m_2} \mathfrak{P}_3^{m_3} \cdots,$$

and determines the two polygons $\mathfrak{A}'_r, \mathfrak{B}'_r$ by the condition

$$d\tilde{\omega}_r = \frac{\mathfrak{A}_r}{\mathfrak{B}} = \frac{\mathfrak{A}'_r}{\mathfrak{B}'_r},$$

and so that they have no common divisor, then the polygons $\mathfrak{B}'_r$ that appear as lower polygons of the differentials are the following:

a) p of them are $\mathfrak{O}$, and the associated differentials $d\tilde{\omega}_r$ are *differentials of the first kind.*

b) Appearing once, each of the lower polygons $\mathfrak{P}_1^2, \mathfrak{P}_1^3, \ldots, \mathfrak{P}_1^{m_1}$ (when $m_1 \geq 2$); $\mathfrak{P}_2^2, \mathfrak{P}_2^3, \ldots, \mathfrak{P}_2^{m_2}; \mathfrak{P}_3^3, \ldots, \mathfrak{P}_3^{m_3}; \ldots$.

The differentials $\tilde{\omega}_r$ corresponding to the lower polygons $\mathfrak{P}^r$ will be denoted by $dt_{(\mathfrak{P}^{r-1})}$ when a sharper distinction is necessary, and will be called *differentials of the second kind.*

c) Finally, each of the products $\mathfrak{P}_1\mathfrak{P}_2, \mathfrak{P}_1\mathfrak{P}_3, \ldots$ (with fixed $\mathfrak{P}_1$) appears once. The associated differentials $d\tilde{\omega}_r$ will be denoted by $d\pi_{(\mathfrak{P}_1,\mathfrak{P}_r)}$ and will be called *differentials of the third kind.*

Each differential $d\tilde{\omega}$ whose lower polygon is $\mathfrak{B}$ can be represented in the form

$$\text{(3)} \qquad d\tilde{\omega} = \sum_r c_r \, d\tilde{\omega}_r$$

with constant coefficients c_r, and this will be called the *normal form* of the differential $d\tilde{\omega}$. Once the individual differentials $d\tilde{\omega}_r$ of a particular kind have been chosen, then the normal form is *unique*, as follows immediately from the linear independence of the differentials $d\tilde{\omega}_r$.

§32. Residues

Summary and Comments

With their powerful algebraic machinery for differentials now in place, Dedekind and Weber go on to prove an algebraic version of the theorem well known from complex analysis: the sum of the residues of an algebraic differential is zero. They reduce it to the case of rational functions by forming the trace.

1. If $d\tilde{\omega}$ is an arbitrary differential in Ω and if $\mathfrak{P}$ is a point that appears m-tuply in the lower polygon $\mathfrak{B}$ of $d\tilde{\omega}$ ($m \geq 0$), then one chooses a variable that is ∞^1 at $\mathfrak{P}$. It is then possible (by §15, 4), and indeed in only one way, to set

$$\text{(1)} \qquad \frac{d\tilde{\omega}}{dz} = a_{m-2}z^{m-2} + a_{m-3}z^{m-3} + \cdots + a_1 z + a_0 + a_{-1}z^{-1} + \eta z^{-2},$$

where the a are constants and η is a function in Ω that is finite at $\mathfrak{P}$. The coefficient $-a_{-1}$ of $-z^{-1}$ in this expression is called the *residue of the differential $d\tilde{\omega}$ with respect to the point* $\mathfrak{P}$. This definition yields the following theorems:

2. The residue with respect to a point $\mathfrak{P}$ is nonzero if and only if $m > 0$, i.e., when the point $\mathfrak{P}$ really appears in the lower polygon of $d\tilde{\omega}$, and hence it is always 0 for differentials of the first kind.

3. The residue of a sum of differentials equals the sum of the residues of the individual differentials.

4. The residue of a *proper* differential is always equal to 0. Because if σ is a function in Ω, and if the b are constants and σ' is a function that is finite at $\mathfrak{P}$ then

$$\sigma = b_m z^m + b_{m-1}z^{m-1} + \cdots + b_1 z + \sigma',$$

and it follows by differentiation of this expression with respect to z, and the fact that $\frac{d\sigma'}{dz}$ is infinitely small of at least second order at $\mathfrak{P}$ (§23, 10), that the az^{-1} term does not appear in the expression for $\frac{d\sigma}{dz}$. This proves the assertion.

5. The residue of a differential $d\tilde{\omega}$ is independent of the choice of variable z. If z_1 is a second variable of the same character as z, i.e.,

$$(2) \qquad\qquad z = az_1 + \zeta,$$

where a is constant and ζ is finite at $\mathfrak{P}$, then it follows, using the abbreviation

$$\alpha = \frac{a_{m-2}z^{m-1}}{m-1} + \frac{a_{m-3}z^{m-2}}{m-2} + \cdots + a_0 z,$$

that

$$\frac{d\tilde{\omega}}{dz_1} = \frac{d\tilde{\omega}}{dz}\frac{dz}{dz_1} = \frac{d\alpha}{dz_1} + a_{-1}z^{-1}\frac{dz}{dz_1} - \eta\frac{dz^{-1}}{dz_1}.$$

Now it follows easily from §23 and §15, 4 that

$$z^{-1}\frac{dz}{dz_1} = z_1^{-1} + z_1^{-2}\zeta', \qquad \frac{dz^{-1}}{dz_1} = z_1^{-2}\zeta'',$$

where ζ', ζ'' are functions that are finite at $\mathfrak{P}$, and the correctness of the assertion then follows by 3, 4.[52]

6. *The sum of the residues of any differential $d\tilde{\omega}$ with respect to all points $\mathfrak{P}$ is always zero.*

To prove this important theorem we can confine ourselves to considering the residues belonging to the distinct points in the lower polygon $\mathfrak{B}$ of $d\tilde{\omega}$. However, we add to these the same number of distinct arbitrary points with vanishing residues until we obtain a polygon $\mathfrak{P}_1\mathfrak{P}_2\ldots\mathfrak{P}_n$ consisting of simple points only and belonging to a proper class. Then we choose a variable z of order n whose lower polygon is just this one, and which is therefore ∞^1 at precisely the points $\mathfrak{P}_1, \mathfrak{P}_2, \ldots, \mathfrak{P}_n$. Among the latter we then have all the distinct points in $\mathfrak{B}$. Under this hypothesis we have

$$(3) \qquad \frac{d\tilde{\omega}}{dz} = a_{m-2}^{(\iota)}z^{m-2} + a_{m-3}^{(\iota)}z^{m-3} + \cdots + a_0^{(\iota)} + a_{-1}^{(\iota)}z^{-1} + \eta^{(\iota)}z^{-2},$$

for $\iota = 1, 2, \ldots, n$, where $\eta^{(\iota)}$ denotes a function that is finite at $\mathfrak{P}_\iota$. If we allow the constants $a^{(\iota)}$ to take the value 0 then the exponent m can be taken to be independent of ι (when not all $a_{m-2}^{(\iota)}$ vanish, m is then the exponent of the highest power of a point appearing in $\mathfrak{B}$). The theorem to be proved now says that $\sum_\iota a_{-1}^{(\iota)} = 0$. In order to prove it, we construct the trace of the function $\frac{d\tilde{\omega}}{dz}$ for the variable z (§2), making use of an extension of the process of §16, 4. We choose a system

---

[52]One can also base the definition of residue on a variable r that is infinitely *small* of first order at $\mathfrak{P}$. Then if

$$\frac{d\tilde{\omega}}{dr} = a_m r^{-m} + \cdots + a_1 r^{-1} + \eta$$

and η is finite at $\mathfrak{P}$, a_1 is the residue of $d\tilde{\omega}$ with respect to $\mathfrak{P}$.

$\rho_1, \rho_2, \ldots, \rho_n$ of functions in Ω as follows: let

$$\rho_1 = 0^m \text{ at } \mathfrak{P}_2, \mathfrak{P}_3, \ldots, \mathfrak{P}_n, \text{ finite and nonzero at } \mathfrak{P}_1,$$
$$\rho_2 = 0^m \text{ at } \mathfrak{P}_1, \mathfrak{P}_3, \ldots, \mathfrak{P}_n, \text{ finite and nonzero at } \mathfrak{P}_2,$$
$$\cdots\cdots\cdots\cdots\cdots\cdots\cdots\cdots\cdots\cdots\cdots\cdots\cdots\cdots\cdots\cdots\cdots\cdots\cdots\cdots$$
$$\rho_n = 0^m \text{ at } \mathfrak{P}_1, \mathfrak{P}_2, \ldots, \mathfrak{P}_{n-1}, \text{ finite and nonzero at } \mathfrak{P}_n,$$

Now if $x_1, x_2, \ldots, x_n$ are rational functions of z, and if

$$\eta = x_1 \rho_1 + x_2 \rho_2 + \cdots + x_n \rho_n$$

is a function in Ω that is finite for $z = \infty$, i.e., at $\mathfrak{P}_1, \mathfrak{P}_2, \ldots, \mathfrak{P}_n$, then $x_1, x_2, \ldots, x_n$ must likewise be finite for $z = \infty$. For if $x_1, x_2, \ldots, x_n$ were *not* all finite for $z = \infty$ there would be a positive exponent r such that the products $x_1 z^{-r}, x_2 z^{-r}, \ldots, x_n z^{-r}$ were all finite for $z = \infty$, with at least one of them, say $x_1 z^{-r}$, nonzero. But then equation

$$\eta z^{-r} = x_2 z^{-r} \rho_1 + \cdots + x_n z^{-r} \rho_n$$

contains the contradiction that at the point $\mathfrak{P}_1$ the left side and all terms but the first on the right-hand side vanish.

It follows immediately, by setting $\eta = 0$, that the functions $\rho_1, \rho_2, \ldots, \rho_n$ form a basis of Ω. Hence if one sets

$$(4) \qquad \frac{d\tilde{\omega}}{dz} \rho_\iota = x_{\iota,1} \rho_1 + x_{\iota,2} \rho_2 + \cdots + x_{\iota,n} \rho_n \qquad (\iota = 1, 2, \ldots, n),$$

where the $x_{\iota,\iota'}$ are rational functions of z, then by §2

$$(5) \qquad Tr\left(\frac{d\tilde{\omega}}{dz}\right) = x_{1,1} + x_{2,2} + \cdots + x_{n,n}.$$

Now it follows from (3) that $z^{-m+2} \frac{d\tilde{\omega}}{dz} \rho_\iota$ is finite for $z = \infty$ and hence so too is

$$z^{-m+2} x_{\iota,\iota'},$$

by the property of the functions ρ just proved. Now, e.g., the functions $\rho_1, \rho_3, \ldots, \rho_n$ are infinitely small of mth order at the point $\mathfrak{P}_2$, while ρ_2 is finite and nonzero there. Hence at $\mathfrak{P}_2$ the functions

$$z \frac{d\tilde{\omega}}{dz} \rho_1, \quad z x_{1,1} \rho_1, \quad z x_{1,3} \rho_3, \quad \ldots, \quad z x_{1,n} \rho_n$$

all vanish, and it follows that $z x_{1,2}$ also vanishes for $z = \infty$. The same follows for $z x_{1,3}, \ldots, z x_{1,n}$ and in general for $z x_{\iota,\iota'}$ as long as ι, ι' are different. Therefore $z^2 x_{\iota,\iota'}$ is finite for $z = \infty$.

If one now sets

$$(6) \qquad x_{\iota,\iota'} = a_{m-2}^{(\iota)} z^{m-2} + a_{m-3}^{(\iota)} z^{m-3} + \cdots + a_{-1}^{(\iota)} z^{-1} + x_\iota z^{-2},$$

where x_ι is a new rational function, then it follows from (3) that

$$x_{\iota,\iota'} - \frac{d\tilde{\omega}}{dz} = z^{-2}(x_\iota - \eta^{(\iota)}),$$

and from (4) that

$$(\eta^{(\iota)} - x_\iota)\rho_\iota = z^2 x_{\iota,1} \rho_1 + \cdots + z^2 x_{\iota,\iota-1} \rho_{\iota-1} + z^2 x_{\iota,\iota+1} \rho_{\iota+1} + \cdots + z^2 x_{\iota,n} \rho_n.$$

Now since $\eta^{(\iota)}$ is finite at $\mathfrak{P}_\iota$ and ρ_ι is nonzero, while all the terms on the right-hand side are zero there, it follows that x_ι is also finite at the point $\mathfrak{P}_\iota$. Since x_ι is rational, this means x_ι is finite for $z = \infty$. It then follows from (5) and (6) that

$$(7) \quad Tr\left(\frac{d\tilde{\omega}}{dz}\right) = \sum_\iota a^{(\iota)}_{m-2}z^{m-2} + \sum_\iota a^{(\iota)}_{m-3}z^{m-3} + \cdots + \sum_\iota a^{(\iota)}_{-1}z^{-1} + \sum_\iota x_\iota z^{-2}.$$

On the other hand, when $\mathfrak{U}$ is once again the lower polygon, and $\mathfrak{Z}$ is the ramification polygon, of z, we have

$$\frac{d\tilde{\omega}}{dz} = \frac{\mathfrak{U}^2\mathfrak{A}}{\mathfrak{Z}\mathfrak{B}},$$

and $\mathfrak{B}$ contains no point that is not also in $\mathfrak{U}$. It then follows as in §23[53] that $\frac{d\tilde{w}}{dz}$, regarded as a function of z, is a function in the module $\mathfrak{e}$ complementary to $\mathfrak{o}$, and consequently

$$Tr\left(\frac{d\tilde{\omega}}{dz}\right)$$

is a *polynomial* function of z (§11, 4). In view of this, it follows from (7) that $\sum_\iota x_\iota = 0$, and also the theorem to be proved:

$$\sum_\iota a^{(\iota)}_{-1} = 0.$$

We can also express these theorems in the following way: the residue of a differential of the second kind $dt_{(\mathfrak{P}^r)}$ with respect to the point $\mathfrak{P}$ is zero.

The residues of a differential of the third kind $d\pi_{(\mathfrak{P}_1,\mathfrak{P}_2)}$ with respect to $\mathfrak{P}_1, \mathfrak{P}_2$ are equal and opposite, and certainly nonzero, otherwise $d\pi$ would be a differential of the first kind.

It follows from these remarks and 4 that a proper differential $d\sigma$ in normal form cannot contain a differential of the third kind. It is also worth mentioning that the residues of the *logarithmic* differential $\frac{d\sigma}{\sigma}$ are integers, namely the order numbers of the function σ (by virtue of §23).

§33. Relations between differentials of the first and second kinds

Summary and Comments

In this section, Dedekind and Weber finally prove a theorem that looks like Abel's theorem, though of course with differentials in place of integrals.

1. Let σ be a function in Ω with lower polygon

$$\mathfrak{B}' = \mathfrak{P}_1^{m_1-1}\mathfrak{P}_2^{m_2-1}\cdots \qquad (m_1, m_2, \ldots \geq 2)$$

and ramification polygon (§16)

$$\mathfrak{S} = \mathfrak{S}'\mathfrak{P}_1^{m_1-1}\mathfrak{P}_2^{m_2-2}\cdots,$$

---

[53]Dedekind and Weber say §26, but §23, 5 seems more relevant, since that is where it is proved that the derivative of a function in $\mathfrak{o}$ is in $\mathfrak{e}$. (Translator's note.)

where $\mathfrak{S}'$ is not divisible by the distinct points $\mathfrak{P}_1, \mathfrak{P}_2, \ldots$. Then in the symbolism of §25 we have the proper differential

$$d\sigma = \frac{\mathfrak{S}}{\mathfrak{B}'^2} = \frac{\mathfrak{S}'}{\mathfrak{P}_1^{m_1} \mathfrak{P}_2^{m_2} \cdots},$$

which shows, in the first place, that *a proper differential can never be of the first kind.*

2. The proper differential $d\sigma$, which in its normal form representation can contain only differentials of the first and second kind, belongs to the vector space of those differentials whose lower polygon is

$$\mathfrak{B} = \mathfrak{P}_1^{m_1} \mathfrak{P}_2^{m_2} \cdots = \mathfrak{B}' \mathfrak{P}_1 \mathfrak{P}_2 \cdots.$$

Conversely, one can always find at least one proper differential $d\sigma$ in such a vector space, assuming that $m_1, m_2, \ldots \geq 2$ and that $\mathfrak{B}'$ belongs to a *proper polygon class.* Because by 1 it is only necessary that a function σ with lower polygon $\mathfrak{B}'$ exist in Ω.

3. This now yields the following important theorem. *All differentials of the second kind may be expressed linearly with constant coefficients in terms of p suitably chosen differentials of the second kind, differentials of the first kind, and proper differentials.*

In order to see this, one chooses an arbitrary polygon $\mathfrak{A}$ of the second kind and order p. Now if $\mathfrak{P}$ is an arbitrary point and r is a positive exponent then the polygon $\mathfrak{A}\mathfrak{P}^r$ is likewise of the second kind, and hence the divisor $\mathfrak{M}$ of the associated class cannot be divisible by $\mathfrak{P}$; otherwise $\mathfrak{A}\mathfrak{P}^{r-1}$, and hence also $\mathfrak{A}$, would be a polygon of the first kind (§30, 4). Therefore, if one sets

$$\mathfrak{A}\mathfrak{P}^r = \mathfrak{M}\mathfrak{B}',$$

then $\mathfrak{P}$ does not divide $\mathfrak{M}$, and hence $\mathfrak{B}'$ contains the factor $\mathfrak{P}$ exactly r times more often than does $\mathfrak{A}$. At the same time, $\mathfrak{B}'$ belongs to a proper class. Now, if

$$\mathfrak{B}' = \mathfrak{P}^{m+r} \mathfrak{P}'^{m'} \mathfrak{P}''^{m''} \cdots,$$

then the powers $\mathfrak{P}^m, \mathfrak{P}'^{m'}, \mathfrak{P}''^{m''}, \ldots$ of points all divide $\mathfrak{A}$. Thus if we set

$$\mathfrak{B} = \mathfrak{P}^{m+r+1} \mathfrak{P}'^{m'+1} \mathfrak{P}''^{m''+1} \cdots = \mathfrak{B}' \mathfrak{P} \mathfrak{P}' \mathfrak{P}'' \cdots,$$

then, by 2, the vector space of differentials associated with the lower polygon $\mathfrak{B}$ certainly contains a proper differential $d\sigma$. The normal form representation of the latter certainly contains the differential

(1) $$dt_{(\mathfrak{P}^{m+r})},$$

and also some or all of the differentials

(2) $$\begin{cases} dt_{(\mathfrak{P})}, dt_{(\mathfrak{P}^2)}, \ldots, dt_{(\mathfrak{P}^m)}, \ldots, dt_{(\mathfrak{P}^{m+r-1})}, \\ dt_{(\mathfrak{P}')}, dt_{(\mathfrak{P}'^2)}, \ldots, dt_{(\mathfrak{P}'^{m'})}, \\ dt_{(\mathfrak{P}'')}, dt_{(\mathfrak{P}''^2)}, \ldots, dt_{(\mathfrak{P}''^{m''})}, \\ \cdots\cdots\cdots\cdots\cdots\cdots\cdots \end{cases}$$

in addition to differentials of the first kind. Thus the differential (1) is expressible linearly with constant coefficients in terms of (2), differentials of the first kind, and $d\sigma$.

Therefore, if the p-gon of the second kind is

$$\mathfrak{A} = \mathfrak{P}_1^{m_1} \mathfrak{P}_2^{m_2} \cdots ,$$

one sees by repeated application of the process just described that all differentials of the second kind are expressible in terms of the p differentials

(3)
$$\begin{cases} dt_{(\mathfrak{P}_1)}, \dots, dt_{(\mathfrak{P}_1^{m_1})}, \\ dt_{(\mathfrak{P}_2)}, \dots, dt_{(\mathfrak{P}_2^{m_2})}, \\ \dots\dots\dots\dots\dots \end{cases}$$

in the manner described in our theorem.

Braunschweig and Königsberg, October 1880.

Bibliography

Abel, N. H. (1826/1841). Mémoire sur une propriété générale d'une classe très étendue de fonctions transcendantes. *Mémoire des Savants Étrangers 7*, 176–264. In his *Œuvres Complètes*, II: 145–211.

Abel, N. H. (1827). Recherches sur les fonctions elliptiques. *J. reine und angew. Math. 2*, 101–181. *3*, 160–190. In his *Œuvres Complètes* 1: 263–388.

Artin, E. (1951). *Algebraic Numbers and Algebraic Functions. I.* Institute for Mathematics and Mechanics, New York University, New York.

Bernoulli, J. (1694). Curvatura laminae elasticae. *Acta. Erud. 13*, 262–276.

Bernoulli, J. (1702). Solution d'un problème concernant le calcul intégral, avec quelques abrégés par raport à ce calcul. *Mém. Acad. Roy. Soc. Paris*, 289–297. In his *Opera Omnia* 1: 393–400.

Bernoulli, J. (1704). Positionum de seriebus infinitis earumque usu in quadraturis spatiorum et rectificatinibus curvarum pars quinta. In his *Werke* 4: 127–147.

Bliss, G. A. (1933). *Algebraic Functions.* American Mathematical Society.

Brill, A. and M. Noether (1874). Über die algebraischen Functionen und ihre Anwendungen in der Geometrie. *Math. Ann. 7*, 269–310.

Cauchy, A.-L. (1844). Mémoires sur les fonctions complémentaires. *Comptes Rendus des Séances de l'Académy des Sciences XIX*, 1377–1384. In his *Œuvres*, Série 1, Tome 8: 378–385.

Clebsch, A. (1865). Ueber diejenigen ebenen Curven, deren Coordinaten rationale Functionen eines Parameters sind. *J. reine angew. Math. 64*, 43–65.

Cohn, P. M. (1991). *Algebraic Numbers and Algebraic Functions.* Chapman and Hall Mathematics Series. London: Chapman & Hall.

Dedekind, I., P. Dugac, W.-D. Geyer, and W. Scharlau (1981). *Richard Dedekind, 1831–1981.* Braunschweig: Friedr. Vieweg & Sohn. Eine Würdigung zu seinem 150. Geburtstag. [An appreciation on the occasion of his 150th birthday.] Edited by Scharlau.

Dedekind, R. (1871). Supplement X. In Dirichlet's *Vorlesungen über Zahlentheorie*, 2nd ed., Vieweg 1871.

Dedekind, R. (1877). *Theory of Algebraic Integers.* Cambridge: Cambridge University Press. Translated from the 1877 French original, with an introduction, by John Stillwell.

Dedekind, R. (1894). Supplement XI. In Dirichlet's *Vorlesungen über Zahlentheorie*, 4th ed., Vieweg 1894. Reprinted by Chelsea 1968.

Dedekind, R. and H. Weber (1882). Theorie der algebraischen Functionen einer Veränderlichen. *J. reine und angew. Math. 92*, 181–290.

Dieudonné, J. (1972). The historical development of algebraic geometry. *Amer. Math. Monthly 79*, 827–866.

Dieudonné, J. (1985). *History of Algebraic Geometry.* Wadsworth Mathematics Series. Belmont, CA: Wadsworth International Group. Translated from the French by Judith D. Sally.

Edwards, H. M. (1980). The genesis of ideal theory. *Arch. Hist. Exact Sci. 23*(4), 321–378.

Eichler, M. (1966). *Introduction to the Theory of Algebraic Numbers and Functions.* Translated from the German by George Striker. New York: Academic Press.

Eisenstein, G. (1847). Beiträge zur Theorie der elliptische Functionen. *J. reine und angew. Math. 35*, 137–274.

Euler, L. (1752). Elementa doctrinae solidorum. *Novi Comm. Acad. Sci. Petrop. 4*, 109–140. In his *Opera Omnia*, ser. 1, 26: 71–93.

Euler, L. (1768). *Institutiones calculi integralis. Opera Omnia*, ser. 1, 11.

Euler, L. (1770). *Elements of Algebra*. Translated from the German by John Hewlett. Reprint of the 1840 edition, with an introduction by C. Truesdell, Springer-Verlag, New York, 1984.

Fagnano, G. C. T. (1718). Metodo per misurare la lemniscata. *Giorn. lett. d'Italia 29*. In his *Opere Matematiche*, 2: 293–313.

Galois, E. (1832/1846). Lettre de Galois à M. Auguste Chevalier. *J. de math. pures et appl. XI*.

Gauss, C. F. (1801). *Disquisitiones arithmeticae*. Translated and with a preface by Arthur A. Clarke. Revised by William C. Waterhouse, Cornelius Greither and A. W. Grootendorst and with a preface by Waterhouse, Springer-Verlag, New York, 1986.

Gauss, C. F. (1832). Theoria residuorum biquadraticorum. *Comm. Soc. Reg. Sci. Gött. Rec. 4*. In his *Werke* 2: 67–148.

Geyer, W.-D. (1981). Die Theorie der algebraischen Funktionen der einer Veränderlichen nach Dedekind und Weber. In Dedekind et al. (1981).

Gray, J. J. (1998). The Riemann-Roch theorem and geometry, 1854–1914. In *Proceedings of the International Congress of Mathematicians, Vol. III (Berlin, 1998)*, pp. 811–822 (electronic).

Hensel, K. (1897). Über eine neue Begründung der Theorie der algebraischen Zahlen. *Jahresber. Deutsch. Math. Verein 6*, 83–88.

Hensel, K. and G. Landsberg (1902). *Theorie der algebraischen Funktionen einer Variablen und ihre Anwendung auf algebraische Kurven und Abelsche Integrale*. Leipzig: Teubner. Reprinted by Chelsea Publishing Co., New York, 1965.

Hilbert, D. (1904). Über das Dirichletsche Prinzip. *Math. Ann. 59*, 161–186. In his *Gesammelte Abhandlungen* III: 15–37.

Hurwitz, A. (1891). Ueber Riemann'sche Flächen mit gegebenen Verzweigungspunkten. *Math. Ann. 39*, 1–60.

Jacobi, C. G. J. (1829). Letter to Legendre, 14 March 1829. In his *Werke* 1: 439.

Klein, F. (1882). *On Riemann's Theory of Algebraic Functions and Their Integrals*. New York, NY: Dover. Translated from the 1882 German original by Frances Hardcastle.

Koch, H. (1991). *Introduction to Classical Mathematics. I*. Dordrecht: Kluwer Academic Publishers Group. Translated and revised from the 1986 German original by John Stillwell.

Koebe, P. (1907). Über die Uniformisierung beliebiger analytischer Kurven. *Göttinger Nachrichten*, 191–210.

Kronecker, L. (1882). Grundzüge einer arithmetischen Theorie der algebraischen Grössen. *J. reine und angew. Mathematik 92*, 1–122. In his *Werke* 2, 237–387.

Kummer, E. E. (1844). De numeris complexis, qui radicibus unitatis et numeris realibus constant. *Gratulationschrift der Univ. Breslau zur Jubelfeier der Univ. Königsberg*. Also in Kummer (1975), vol. 1, 165–192.

Kummer, E. E. (1975). *Collected Papers*. Berlin: Springer-Verlag. Volume I: Contributions to Number Theory, edited and with an introduction by André Weil.

Lemmermeyer, F. (2009). Jacobi and Kummer's ideal numbers. *Abh. Math. Semin. Univ. Hambg. 79*(2), 165–187.

Lüroth, J. (1875). Beweis eines Satzes über rationale curven. *Math. Ann. 9*, 163–165.

Lützen, J. (1990). *Joseph Liouville 1809–1882: Master of Pure and Applied Mathematics*, Volume 15 of *Studies in the History of Mathematics and Physical Sciences*. New York: Springer-Verlag.

Mittag-Leffler, G. (1923). Preface. *Acta. Math. 39*, i–iv.

Möbius, A. F. (1863). Theorie der Elementaren Verwandtschaft. In his *Werke* 2: 433–471.

Neumann, C. (1865). *Vorlesungen über Riemann's Theorie der Abelschen Integralen*. Leipzig: Teubner.

Poincaré, H. (1882). Théorie des groupes fuchsiens. *Acta Math. 1*, 1–62. In his *Œuvres* 2: 108–168. English translation in Poincaré (1985), 55–127.

Poincaré, H. (1883). Mémoire sur les groupes Kleinéens. *Acta Math. 3*, 49–92. English translation in Poincaré (1985), 255–304.

Poincaré, H. (1907). Sur l'uniformisation des fonctions analytiques. *Acta Math. 31*, 1–63. In his *Œuvres* 4: 70–139.

Poincaré, H. (1918). *Science et Méthode*. Paris: Flammarion. English translation by Bruce Halsted in *The Foundations of Science*, Science Press, New York, 1929, 357–553.

Poincaré, H. (1985). *Papers on Fuchsian Functions*. New York: Springer-Verlag. Translated from the French and with an introduction by John Stillwell.

Riemann, G. F. B. (1851). Grundlagen für eine allgemeine Theorie der Functionen einer veränderlichen complexen Grösse. In his *Werke*, 2nd ed., 3–48.

Riemann, G. F. B. (1857). Theorie der Abel'schen Functionen. *J. reine und angew. Math. 54*, 115–155. In his *Werke*, 2nd ed., 82–142.

Riemann, G. F. B. (2004). *Collected Papers*. Kendrick Press, Heber City, UT. Translated from the 1892 German edition by Roger Baker, Charles Christenson and Henry Orde.

Roch, G. (1865). Ueber die Anzahl der willkürlichen Constanten in algebraischen Functionen. *J. reine und angew. Math. 64*, 372–376.

Salmon, G. (1851). Théorèmes sur les courbes de troisième degré. *J. reine und angew. Math. 42*, 274–276.

Schwarz, H. A. (1872). Über diejenigen Fälle, in welchen die Gaussische hypergeometrische Reihe eine algebraische Function ihres vierten Elementes darstellt. *J. reine und angew. Math. 75*, 292–335. In his *Mathematische Abhandlungen* 2: 211–259.

Shafarevich, I. R. (1994). *Basic Algebraic Geometry. 1* (second ed.). Berlin: Springer-Verlag. Translated from the 1988 Russian edition and with notes by Miles Reid.

Smithies, F. (1997). *Cauchy and the Creation of Complex Function Theory*. Cambridge: Cambridge University Press.

Stevin, S. (1585). *L'arithmetique*. Abridgement in *Principal Works of Simon Stevin*, vol. IIB, 477–708.

Walker, R. J. (1950). *Algebraic Curves*. Princeton Mathematical Series, vol. 13. Princeton, NJ: Princeton University Press.

Weber, H. M. (1908). *Algebra, Band III*. Braunschweig: Vieweg. Reprinted by Chelsea, 1979.

Weierstrass, K. (1863). Vorlesungen über die Theorie der elliptischen Funktionen. *Mathematische Werke* 5.

Weierstrass, K. (1870). Über das sogenannte Dirichlet'sche Prinzip. Read in the Berlin Academy, 14 July 1870; in his *Werke* 2: 49–54.

Weil, A. (1975). Introduction to Kummer (1975).

Weyl, H. (1913). *Die Idee der Riemannschen Fläche*, Volume 5 of *Teubner-Archiv zur Mathematik. Supplement [Teubner Archive on Mathematics. Supplement]*. Stuttgart: B. G. Teubner Verlagsgesellschaft mbH. Reprint of the 1913 German original, with essays by Reinhold Remmert, Michael Schneider, Stefan Hildebrandt, Klaus Hulek and Samuel Patterson. Edited and with a preface and a biography of Weyl by Remmert.

Weyl, H. (1964). *The Concept of a Riemann Surface*. Translated from the third German edition by Gerald R. MacLane. ADIWES International Series in Mathematics. Addison-Wesley Publishing Co., Inc., Reading, Mass.-London.

Index

Abel, N. H., vii
  and elliptic functions, 4, 16
  and genus, 5, 8
  constraints on existence
    for elliptic functions, 15
  paper on Abel's theorem, 6
  theorem of, 2, 5, 41, 138
    Dedekind-Weber version, 123, 125, 138
Abelian differential, 121, 122
Abelian integral, 41, 43
Acta Mathematica, 20
algebraic curve
  and calculus, 2
  as covering of the sphere, 8
  as Riemann surface, 6, 12, 14, 28
  functions on, 14
  genus of, 6
  integral on, 12
  parameterization, 20
  with no rational parameterization, 29
algebraic function, 1, 2, 41
  field, 27, 42, 45
    basis, 47
    degree of, 46
  integral, 32
  norm, 32, 48
  of bounded degree, 14
  over algebraic numbers, 42
algebraic integer, 24
  Dedekind definition, 25
  norm of, 25
algebraic numbers, 24, 41
  and ideal theory, 1
arithmetization, 37
automorphic function, 20

base locus, 107
basis
  complementary, 75, 76, 123, 124
  for ring of integral algebraic functions, 54
  modulo a module, 60
  normal, 110, 112, 124
  of algebraic function field, 47
  of integral algebraic functions, 53

of module, 55
of polygon class, 109
of polynomials, 77
of vector space, 59
  of polygons, 107
Bernoulli, Jakob, 3
  and lemniscatic integral, 3, 4
  nonrational curve, 29
Bernoulli, Johann, 3
birational equivalence, 17, 29
  and isomorphic function fields, 29, 31
  of algebraic curves, 29
  of curves of genus 1, 31
  of sphere with any genus 0 surface, 29
birational geometry, 6
birational transformation, 89
Birkhoff, G.D., 110
Bolyai, J., 18
branch point, 7, *see also* ramification point
branching, 7, *see also* ramification

canonical class, 15, 123, 126, 127
  is proper, 127, 130
Cauchy, A.-L., 6
  and meromorphic functions, 14
  and Liouville's theorem, 13
  integral formula, 12
  integral theorem, 12
  residue theorem, 12
  theory of integration, 12
chain rule, 114
characteristic polynomial, 60
Chevalier, A., 6
class
  canonical, 126
  polygon, 105
  principal, 123, 126, 127
Clebsch, R. F. A, 8, 41
compactness, 14, 33
  simplifies meromorphic functions, 14
complementary
  basis, 75, 76, 123, 124
  module, 75, 80, 117, 124, 138
complete system of remainders, 59, 60

145